FUNDAMENTALS
OF
PROGRAMMABLE LOGIC CONTROLLERS
AND
Ladder Logic

Orlando Charria
First Edition

Theory and Hands-On
40 Step by Step Lessons
14 Automation Projects

Published by Latin Tech Inc.
Miami, Florida.

ISBN 10:   0615800076
ISBN 13:   9780615800073

To my son David
for making my life worthwhile

IMPORTANT

Get your demo software now!

In order to get a DEMO license to practice the ladder logic lessons and projects, please send an email to : **plctraining@hotmail.com** and **sales@latin-tech.net** including your info and the serial/receipt number of the book you have bought.

We really want you to learn so you can use above emails to contacts us for:
- General doubts.
- Help on lessons and projects.
- Help on installing the software.
- Help on writing your first program and using the software (Simulation and downloading).

There is no charge for these services. We use free remote access tools to provide assistance.

Just in case that you want to practice with a real PLC, check our PLC trainer on page 403.

We look for positive reviews

FROM THE AUTHOR

Just a few words

Being a lecturer and having performed several trainings on industrial Programmable Automation Controllers has thought me that if you really want to learn about PLCs you have to split the process in two: First you should master the theory and second, you must practice as much as you can.

When I first wanted to write this book I was really enthusiastic thinking that it would be a very good idea to cover all the topics involved in an automation project. I even made a list of everything I wanted to write about, but perhaps my emotions were so extreme, that the list resulted in several pages.

I decided to write about Ladder Programming and I tried to make a new list but unfortunately I ended up with similar results.

I thought to myself: "Maybe I want to teach much information" and after re-analyzing everything, instead of removing topics from my list, I ended up adding many more. I also remembered that the main problem with most of the books in the market is that they are just theory or examples but not both in the same content.

I didn't want to fall into the problem that a wise saying reflects: "An Ocean of wisdom with a centimeter of depth" This was the origin of the book and the support for the main goal: Introducing just some basic automation concepts, the PLC with its programming techniques, then some lessons and finally adding as many practical examples as possible.

Since this is a book about fundamentals, the instructions were restricted just to the ones considered as basic and common to all PLCs.

If you are using this book with any of our PLC trainers then, in addition to the simulation, you can download the programs and examples to the PLC and experience the real feeling of an automation project.

Since the PLC trainer has pushbuttons, lights, relays, etc., then you can realize that the automation environment becomes real for you.

You can even make connections of different devices to the trainer.

Whether you just have the book or the book with the trainer I really want to congratulate you on this first step to gain knowledge about an exciting world. I hope you enjoy this book as much as I did when I wrote it.

For the journey you have just started ... there won't be limits anymore.

Orlando Charria.

ABOUT THE AUTHOR

Mr Charria holds an Electronics Engineering degree from Universidad Pontificia Bolivariana in his natal country Colombia. He also has some post-graduate studies and an Individual Program in Holland in the field of Machine Vision and also in the field of environmental sciences.

He has more than twenty five years working with all type of industrial automation products and processes and serving as lecturer and professor in many educational institutions in Colombia and USA. He has also participated as an engineer and consultant in automation projects all over the world and helped several OEMs to define the automation strategy for their machinery.

His experience as engineer relates to very important companies like Philips, VARTA and more recently, Latin-Tech.

CONTENTS

INTRODUCTION

1

The number of automation devices, installed in the whole world, is constantly increasing. We could think that most of the known industrial processes are using at least one PLC to operate.

Since this knowledge is evolving throughout the time it has become a real need for the technicians to update their skills on how to wire, troubleshoot, program and test all PLCs and the machines they control.

In the market you will find a lot of really good books about Automation, PLCs, ladder programming and programming techniques. But the main problem is that the contents are either very specific or very specialized. For instance: A book dedicated to ladder programming will explain the methodologies but will assume that you know a lot about automation.

If the book is about automation, it will cover all general aspects, but most of the chapters related to ladder programming will be summarized and the examples will be limited.

This book is intended to cover fundamentals on PLCs and ladder programming so it would be like you were buying three different books:

The first book helps you to understand the fundamentals on automation and ladder logic.

The second book covers the basic aspects of ladder logic, including several lessons for you to master the use of the basic instructions.

A third book about real PLC applications where you can have real hands-on programming and testing through the included simulation and programming software or making real connections to our PLC trainers.

This book is the perfect companion for the training equipment we do offer.

In addition to the knowledge you can get with the Software simulator, you can also develop real projects using the PLC trainer.

Keep in mind that to learn and understand the concepts, the theory must be accompanied by as much practice as possible.

Remember, it´s the practice that will turn you into a master.

ORIGIN OF PLCs

2

[This is a short tribute to the inventor of the
Programmable Logic Controllers]

Dick Morley, Father of PLCs

Since the industrial automation is in permanent change, a lot
of information is written everyday... always pointing to the
future. But just a few of the articles try to honor the past and
the origin of this technology when the first PLC was conceived.

The first PLC, Model 084, was invented by Dick Morley in 1969.

The first successful commercial PLC, the 184, was introduced
in 1973 and was designed by Michael Greenberg.

These words are written to reconstruct the birth of the PLC and
to thank the creator, because without his first idea this world
of automation would not be what it's today: The invention
or innovation that transformed the industry, generated a $4
billion dollar market, and changed the scope of any control
process in terms of quality, reliability and flexibility.

Mr. Richard (Dick) Morley is considered the father of the programmable controller. For about 30 years he has been involved in a lot of high tech companies. He was part of the team who developed the floppy disk. With several patents in USA and abroad, his visionary work has contributed the growth of technology.

Because of his innovation in PLC, in 1997 the Franklin Institute honored him the Prometheus Medal. He holds third place of the International IEN among the "Top 100 Most Significant Industrial Products of the 20th Century." In October 1999, the famous Instrumentation, Systems and Automation Society (ISA) gave him the "Life Achievement Award" . Five months later, in March, Fortune magazine also awarded him with the "Heroes of Manufacturing Award".

Dick Morley graduated in Physics from the Massachusetts Institute of Technology in 1954. Nothing has ever limited his growing achievements. He later attended Northeastern University [Boston] for about a year, not to get a degree but to complement some background he was missing.

For more than 50 years his life has been really prolific since he has been engineer, author, consultant, inventor and an angel investor for more than 100 new technology based companies.

Questions.

1. Research for the "moment" where the idea of a PLC appeared for the very first time.
2. Browse the internet to look for new ideas Mr. Morley has been working on.

¿WHAT IS A PLC?

3

This chapter teaches the way to identify the definition of a PLC, its advantages and all the possible application fields.

The term PLC stands for Programmable Logic Controller and can be defined as a control device which can be used to provide accurate, reliable and continuous operation of control tasks.

A PLC is an electronic equipment that :

- Helps to have a more continuous operation for a machine or process.

- Drastically reduces the maintenance.

- Helps to protect the life of the operators.

- Simplifies the diagnostic or troubleshooting.

- Guarantees repetitive operational conditions.

- Simplifies the labor for the operator.

- Guarantees an efficient and cost effective process.

- Protects the life of the control elements involved.

- Helps to get a high quality product.

- Enhances the production level.

- Raises competitiveness of the companies.

- Aids in the acquisition of current online data.

The electronic design of the PLCs must be made to support the most demanding electrical and mechanical operational conditions such as voltage variations, high temperature, high relative humidity, extreme vibration, etc.

To accomplish such hard specifications, PLC manufacturers developed technologies that utilize components which, because of their purpose, become more useful. In this race to have a better, more reliable and extreme quality product the result has been a lot of different brands that go to the market with all the required aspects for a technological product:

- Cost.

- Service.

- Better technical specifications.

- Market niche.

- Operational environments.

- Available options.

- Connectivity.

A PLC can be used in all types of applications:

- Industrial Machinery.

- Home and Building automation.

- Robotics.

- Automated vehicle guidance.

- Conveyor belts.

- Plastic injection machines.

- Numeric Control for Tool machinery.

- Automated storage in Cold rooms.

- Air conditioning systems.

- Packing machinery.

- Redundancy systems in processes.

- Water and Energy management systems.

- Industrial Waste and Water treatment plants.

- Substations, Energy distribution and transportation systems.

- Parking systems.

- Irrigation systems.

- Car assembly.

- Bottle filling and sealing.

- Amusement parks.

- Automatic lubrication systems.

- Traffic lights.

- Emergency evacuation systems.

- Security systems.

- Vending machines.

- Light and dimmer control.

- Mobile robotics.

For many years, machine manufactures were used to wire all the control logic using the contacts of relays and their respective coils. This situation led to a very complex and large wiring which was demanding in time and maintenance. Trying to locate the reason for a faulty operation was really a difficult task, since the diagnostic tools were scarce.

This kind wiring was called Relay Ladder Logic (or RLL) mainly because it was based on relays and contacts to provide any control.

One of the most important remarks was the wire marking technology as one of the most widely used tool when a problem was detected.

With time, technology on computers brought a new approach.

New processing systems and products appeared to drastically produce changes in control technology which quickly overcame most of the typical problems:

1. Many devices involved large and complex control panels.
2. Just wiring the control required a lot of work.
3. It was difficult to trace a failure in the system.
4. To make simple changes in the control was expensive and difficult.
5. Changes had to be made locally, so OEMs and customers had the problem for a costly support.

The microprocessors and its latest descendant, the

microcontrollers, offered the control designers new possibilities to make faster design and easier to maintain control systems.

Very soon, most of the manufacturers incorporated microprocessors in their PLCs, achieving the following advantages:

1. Control equipment with smaller size and less cost.
2. Programming software tools for fast development.
3. High flexibility to perform changes in control with minimum or no impact on the wiring.
4. Highly powerful diagnostic tools.
5. Support for modems, gave them the possibility to change or monitor the control program from any place in the world.
6. Addition of more powerful and sophisticated instruction set to perform a lot of calculations and many difficult tasks.

The use of PLCs is very popular nowadays. Computer manufacturers have been trying to get a piece of the industrial automation market, but because of the cost and the industrial characteristic, their final cost is still very high.

Miniature electronic components and the fierce competition of computer technology have turned them into modern PCs with more possibility of surviving industrial environments, smaller sizes and lower costs. Some PLC manufacturers are adopting this technology to design new products which can use the best of both worlds: PLCs and PCs.

Control panel with a lot
of wiring and Relays
   VS
Control panel with
PLCs

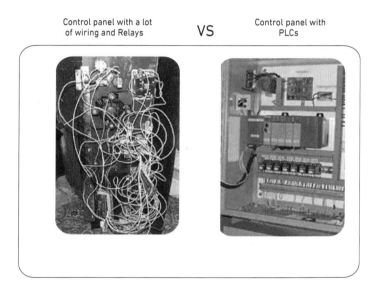

Questions

1. What does "PLC "stand for?
2. When would you use a PLC instead of wired logic?
3. Mention three advantages of using PLCs.
4. Mention three applications of PLCs.
5. Provide a list of industries where the use of PLC is very common.

THREE EXAMPLES OF AUTOMATION

4

A good way to understand all the automation concepts and the importance of this work is by looking to some examples.

4.1 Cookies' Production Plant.

Let´s assume that a company named COOKIE CO is a company which produces a tasty and crunchy cookie originated from one of the grandmother's famous recipes.

4.1.1 The Manual Process.

The production is almost as the original way grandmother used to prepare the cookies. First, all the ingredients are mixed in a big bowl then, using a cup as a measurement tool, the material is deposited on a tray.

A pre-heated oven is used to cook the cookies. After some hours a delicious tray of cookies is ready .

The cookies became really famous and most of the neighbors were interested in buying some. Based on this growing possibility, the grandmother decided that it was time to earn some money and make some profit from her recipe.

The popularity of the cookies was increasing and more orders were coming. The big problem was that this situation was requiring a lot of work from grandmother, leaving her really exhaust at the end of each day.

More income was implying more work...

4.1.2 The Second Step.

Grandmother was becoming very wealthy, but her health was becoming a concern among the relatives. Although they helped her in their free time it wasn't enough to attend the growing demand for grandmother's cookies.

After a family meeting they all agreed in "industrializing" the cookie production. The resulting ideas were to get a bigger oven to allocate more trays so more cookies could be produced and more importantly, getting a full time helper or assistant.

After answering the bank requirements to proof that the money was not required at all(which is what bankers want to know for a bank loan), Grandmother had enough money to have a better company (the author doesn't believe in the way banks do business).

Now she´s got a big oven with an eight-tray capacity and with a temperature control. They believed that the person hired to help grandmother had been a good temporary solution since Grandmother could rest more and avoid unnecessary heavy workloads.

4.1.3 The Manual Industrial Production.

A more industrialized production got started. The orders for Grandma's cookies were increasing. One of the factors to influenced people in buying the cookies was the aroma the factory was spreading all over the neighborhood.

But another problem came into notice. Though Grandma tried to keep track of all the steps during the production and the helper was doing his best some problems appeared:

1. The size for most of the cookies was not equal.

2. The brown color was not uniform during the production.

3. The taste was not as the original ones.

Grandma didn't want to lose her customers so she decided to look for technical advice with her Grandson, who was finishing an engineering education. She asked him to diagnose the problem and provide a good solution.

After analyzing everything, they both agreed on something:the solution was to implement some process adjustments in order to fix the variables that could be depending on a person´s criteria.

They introduced a material feeder system to allow the helper to always pour the same quantity of mass. They defined the exact temperature set-point to heat the oven and established a new policy realted with the final order of when to take the cookies out of the oven: It had to be Grandma´s decision.

With these solutions they sorted out most of the problems, at least for a while. They, somehow ,discovered that the best approach was controlling the variables that could affect the quality of the cookies.

4.1.4 The First Industrial Automation.

Very soon Grandma´s business flourished with more orders, more employees and more equipment (since more ovens had been bought). They had to move to another facility with industrial type installations. Very importantly, they succeeded in paying the bank back all the loan and since the profit was really good and the expenses were in control, the rest of the machinery was bought in cash! (Lucky them!).

The company had generated several jobs and, as a result of this, the town was having a lot of prosperity since the economy of the families was getting better.

The factory now had three production lines. Each line was formed by an automatic mixer, a mass dispenser and a conveyor belt which transported the cookies through an oven.

Since the future plans are to export, they will have to analyze what to do to be competitive, profitable and to produce with high quality standards.

4.2   Refrigerated Storage

The owner of a cold room tried to maintain all kinds of perishable goods in his location, but from time to timesome problems were generating serious economic losses.

The problems could be summarized in:

- The temperature was not fully homogeneous in the whole room. Some products were having a different temperature than others.

- At nights, when the system was unattended, very expensive damages could occur after a power outage.

- The workers used to forget to close the main door which resulted on a drastical temperature change.

- Some external customers, having their perishable products in the cold room, were concerned about the real temperature which their products were exposed to.

Since quality of any perishable product is affected by temperature changes, all the problems, at the end, became an economic matter for the owner. His own products were affected and the same for all the confidence on the services he was providing to the external customers.

Then, he decided to look for a solution to avoid most of troubles by hiring an external technician to develop an automation on the system.

The technician came up with the following independent solutions:

- Non homogeneous temperature.

Two flat fans, located below the ceiling floor, would help to spread the cold around the room. Four temperature sensors installed at different locations, would provide information of the temperature to a control system. The internal program could read the temperature and, when the calculated difference between all of them were only 0.5C it would be considered as "homogeneous" turning the fans off. This condition would save energy by avoiding to have the system working all the time. In general, the fans would be on as long as the temperature is not homogeneous.

-Power outages or Shut-downs.

The energy supply service was operating regularly. Just a few times the energy was disconnected for more than one hour, but in general the power outages were of less than half an hour. This time, however, was more than enough for the room temperature to increase.

A back-up power generator could be used to supply the energy while the power outage was taking place. The big problem was the cost. Adding a system to perform an automatic energy transfer (without human intervention) to switch betteen the utility and the generator, could be very expensive. The technician knew that the right thing to do was to consider all possible options for the owner to make a good decision.

-Temperature tracking system.

In order to help on the service quality, it´s a good idea to design a system to allow both, the owner and his customers, to keep online tracking of the current temperature in the room.

4.3   Building automation.

The maintenance engineer of a big commercial building is noticing that the power bill is increasing. The operating costs are worrying the manager.

The new bill is about 35% more than the last year one.

After a meeting, where most of the directive staff was present, they came to the following conclusions:

- The number of workers has increased a 10%.

- The overtime hours have been reduced .

- The building occupancy is about 60% and almost constant.

-Most of the power is consumed by the Air con-ditioning and the total lighting system.

The maintenance engineer was urged to provide a solution and, as usual in the manufacturing industry, maintenance is fully responsible for what happens. You know, most of the people think: "When things go well maintenance is not needed and when things go bad it's because there is no maintenance at all".

The first thing to do, according to the request, was to know the equipment that was consuming huge amounts of energy.

After a quick survey the maintenance engineer discovered that the Air conditioning, the interior lighting and the outdoor lamps, could be the equipment that was strongly affecting the power bill.

Probably, the best way to try to provide a solution was to understand how every equipment was utilized in order to determine how it´s contributing to the power bill (the amount of money which is spent in operation).

4.3.1 Air Conditioning.

The equipment was a central unit which was controlling the temperature of the whole building. He found that this AC was very old, because it was installed ever since the company started so, after all these years the unit must not be efficient nor in good condition.

Our technician realized that every floor had different number of people. The cooling requirements were different and had t be calculated according to the occupancy.

There were rooms only used for some hours. For instance, the conference room was used just two or three times a week, however the AC was running all the time.

4.3.2 Inner lighting.

All the floors had all the lights on, including places where there were no people.

Most of the employees didn't pay any attention to turn off the light before leaving their offices.

Some places had excessive illumination: this means that there were more lamps than the required.

4.3.3 Outdoor lamps.

The outdoor lighting was also redundant. There were lot of lamps pointing the light to the same location and leaving some other places without illumination. Since the installation was old probably the kind of lamps used was not as efficient as supposed.

4.4 The automation strategy.

The maintenance engineer prepared the following plan in order to get a quick approval for the project:

- Detailed report of things to do that requires investment and the things that can be made without any major cost.

- Definition of priorities.

- Partial cost and total costs.

- Return of investments (ROI) calculations.

- Meeting with the decision making people.

In general, the automation proposal was:

- Making a deep analysis of the floor occupancy. All those floors which could be closed by moving the people to another floor would benefit the costs in the energy bill.

- Cost study to help in the selection of using the same AC unit, having new equipment with higher efficiency or independent circuits and smaller AC units.

- Separate low occupancy rooms (like the conference room) from the main unit.

- Installation of motion sensors in most of the offices to detect the presence of persons at night in order to have a control to turn off the light during the night when there are no persons working.

- Distribution of sensors at given doors to detect if the room has people inside as an alternative of low cost detection versus the motion sensors.

- Addition of a manual way for connection and disconnection of loads. Security personnel for instance, every night can turn on or off the different circuits of light in each floor.

- Installation of equipment to individually measure power and energy per electric circuit. For instance, building floors, lights in offices, computers, high consumption equipment, etc.

- Illumination study, according to norms ,to detect which places are lacking of light or which places have excessive illumination.

- New electrical circuits and lamps for the outdoor lighting systems. This new conception will help to turn on or off just the lights needed, saving a lot of energy.

To make this story short, the maintenance engineer presented this report. Everyone in the meeting was convinced of the solution so, the automation was approved without any hesitation.

The savings were higher than expected so, the maintenance engineer received a substantial bonus at the end of the year.

His budget for automation was raised after seeing the results.

Questions

1. Reviewing all three cases, which can be made in less time?

2. Which case can provide more benefit to the company?

3. Which case would take less time to implement?

4. Could you think about some other examples of automation?

5. Which project has production risks?

6. How do you detect if the automation is not working?

7. How many people are involved in the automation process?

8. Think of the kind of alarms we can add to every project.

9. Try to imagine possible situations in the process that have not been considered here.

10.If you were to develop any of these cases as a commercial project, what are some potential names?

TYPES OF PLCS

5

There are a lot of PLC manufacturers for a very competitive market. To face competition, new PLCs contain relevant features that can separate them from the others. For this technological battle, the PLCs have been reformulated several times in such a way that they can be differentiated in terms of cost, size, quality, reliability, connection, ports,etc.

There are several ways to classify a PLC but the most popular can be based on:

a) Architecture :

- Fixed.

- Modular.

- Mixed.

Fixed   Modular   Mixed

When a PLC has no possibility of expansion we call it of a fixed configuration. It means that the main parts such as the CPU, the Inputs and the Outputs are going to be defined by the model itself and that you won't be able to add more parts.

In Modular PLCs, every part of the PLC is considered as an independent module. You can therefore buy, for instance, inputs modules with certain input capacity and then, at the moment of needing more inputs you have two options: either removing the current input module to put another with more inputs or use extra space in the rack or base and add an additional input module. This is why, in Modular PLCs, you need to be very careful when selecting the size of racks to allocate your modules. As a general rule, you need to leave some unused slots to prevent future I/O growth.

When the PLC already has some inputs and outputs but also has some empty slots, for you to add extra modules, we say that this PLC is a Mixed type.

b) Assembly:

    - Board level.   - Chassis level.

Since most of the time, there is a Control panel or cabinet involved in an automation application, some manufacturers consider that there is no need to protect the PLC so they provide the bare Printed Circuit Board.

Most users rely on the features and technical specifications rather than the external appearance of a given PLC.

Using an external chassis adds extra costs to the manufacturing of the PLC that can be significantly important to users like the so called Original Export Manufacturers (OEMs).

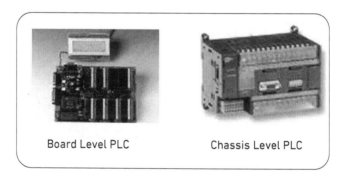

Board Level PLC   Chassis Level PLC

Some other manufacturers consider the external case as part of the branding and marketing strategy, which helps the user to identify the PLC among others. This aesthetic operation has a cost that in most cases both: manufacturers and users, are willing to pay.

b) Power Supply:

The Power supply can be External or Built-in.

All the PLCs require DC voltage to operate their internal circuitry, but the Power utilities supply an AC voltage for any industrial environment. The manufacturers have developed the power supplies to provide a safe DC Voltage to the PLCs and control electronics, implementing designs based on either Voltage transformer and a rectifier or more efficient and sohisticated systems called Switching power supplies.

According to the cost and PLC specifications, you can have the power supply included in one of the parts (For instance the rack or the CPU) or you need to buy an external power supply.

c) Reparability.

Most of the PLCs are not fixable. No matter how reliable they are, as soon as one of the parts is damaged in one of the parts it has to be replaced immediately, otherwise the production they are controlling can have problems and therefore, leading to a costly stop time.

To fix a PLC you are required to order a module or part from the manufacturer, because not always it's possible to carry an inventory of all the possible PLC parts. In some countries to get a new module it's a few hours away, but in other countries the import process could last months. Based on the last situation, some manufacturers have mounted on base all the integrated circuits of every module, to allow quick removal of chips.

This reparability led to another problem because of warranties.

Questions

1. What is a PLC of Fixed I/0?
2. What is a PLC of Modular I/0?
3. What is a PLC of Mixed I/0?
4. What is a PLC in board?
5. What is a PLC with chassis?
6. What is a PLC with built-in power supply?
7. How can you repair a PLC?
8. What is an OEM?
9. What type of PLC is recommended for small applications?
10. What type of PLC is recommended for applications that will be assembled in stages (the PLC will need to grow).
11. When do you apply a mixed architecture PLC?

BASIC ARCHITECTURE OF A PLC APPLICATION

6

This chapter will provide a very updated approach to all the elements involved in an automation project. All new trends and technologies are mentioned.

An automation project requires several devices that interact together to produce a realiable control operation. With the system components you have to develope an automation task which must be built around the control core, in this case the PLC.

In general, an automation project might include at least one PLC and any (or all) of the following components:

1)   Power supply.
2)   PLC.
3)   Operator Panel.
4)   Sensors.
5)   Transmitters.
6)   Interface Elements.
7)   Final Control Element.
8)   Communication devices.
9)   PC.

10)   Universal modules.
11)   Remote Modules.
12)   Hubs.
13)   Drives.
14)   Voice Modules.

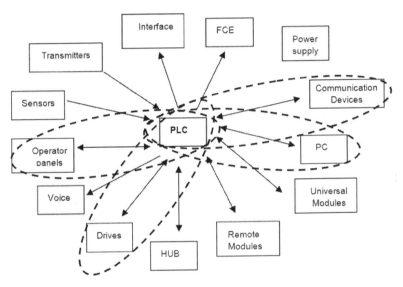

Elements for automation with PLCs

We could write a complete book of several pages trying to explain all the aspects of every component of an automation project. Every device is a complex industry that involves technology, know-how, marketing, accuracy, etc.

We don't intend to provide a detailed coverage of each automation product, but at least a quick view for you to understand some very common automation situations.

To understand the above graphic please take into consideration the following concepts:

1. The rectangles represent devices and a complete world of several possibilities. For instance: Drives include all types of commercial drives: AC, DC, Stepper, Servo, etc.

2. The Arrows represent the direction of the information, hence a two headed arrow means bi-directional information and a single headed arrow means one single direction.

3. Dotted lines represent typical commercial combinations. This means that some manufacturers are joining two different compatible products and creating a new product with powerful features. This is the competition arena.

Now let's try to concentrate on the different components:

6.1   Power supply.

The power supply is common to all the rest of the elements; this is why it has no connection.

The automation devices require a power supply to operate. It provides the specific voltage and current the system requires. Generally, only one power suuply is required for the whole automation project, but if there are some devices with different specifications or out of the standard ranges, more power supplies might be required.

Some power supplies include extra capabilities like: short circuit protection, overvoltage protection, filtering of undesired electrical signals, etc.

You could imagine that there are a lot of types of power supplies, but we will only consider those ones that are really used in Industrial automation.

Lately, new technologies are allowing to have an Uninterrupted Power Supply or UPS. These systems can have a battery to provide the power for a given time in case of having an AC power outage.

There are two main types of power supplies:

Linear and Switching

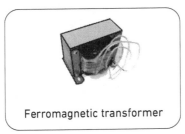

Ferromagnetic transformer

The linear power supplies use a transformer as one of the main devices. The picture below shows the kind of transformer we are talking about.

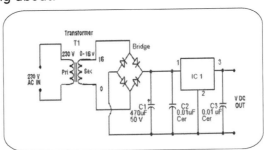

Typical linear power supply diagram.

A linear power supply for industrial automation is formed by three basic stages:

Voltage transformation, Rectifying and Output Regulation.

The power supplies that use transformers have simple electrical diagrams as the one found in the former page. It includes just few components.

The transformers are heavy, bulky and dissipate heat when they are working. This situation puts them in a very non- efficient position when compared to the ones used in switching power supplies.

Today, machine manufacturers prefer to use switching power supplies instead of linear power supplies because of the following advantages:

- Less heat involved so they are more efficient.

- They can absorb any input voltage variation within a very wide range. For instance, they can accept any voltage ranging from 80 to 240VAC.

- Less maintenance required.

- Better output voltage regulation.

- Minimal space required.

- No fuses required since the electronics discon- nect the load when short circuits are present.

- Complex circuitry with better operational featu- res.

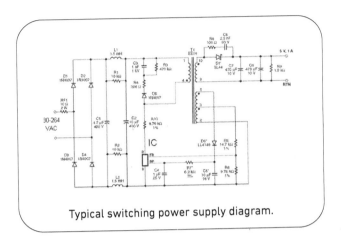

Typical switching power supply diagram.

The figure shows the electric diagram of a very common switching power supply. At first sight you notice that there are more components involved. At this moment is not that important to know about the electronics involved.

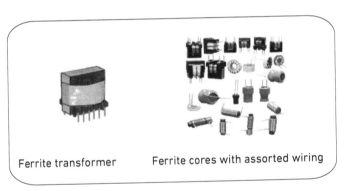

Ferrite transformer   Ferrite cores with assorted wiring

Two components are very important here: The IC, which internally is a very complex device, and the ferrite transformer. This last one is a very small transformer that can work at higher frequencies than the normal transformers.

Switching power supplies.

The above figures, show a ferrite transformer (left) and assorted types of transformers and coils made on ferrite cores (right).

This is the type of power supply the PLC manufacturers include as an internal option, when offering the PLC with built- in power supply.

6.2   PLC.

If you observe the main graphic you'll see that the PLC is like the "center" of everything. This was done on purpose, since the PLC can be considered as the core of the automation solution. There are some other applications where they can be based on PCs, but they do not necessarily use ladder logic which is one the main goals of this book.

The PLC interconnects most of the devices in several ways. Normally, the information from the automation project is collected, processed and controlled by the PLC. Then the PLC is consulted from other external equipment like PCs, but it's good to clarify that the main purpose of the PLC is to control the process.

6.3   Operator Panel.

The interface between the process or machine (controlled by the PLC) and the person who operates the machine is called

Human machine Interface or HMI (We still consider that humans are operating the machines!).

The selction of the right operator panel is very important in order to have an easy, efficient and friendly operation of the machine or process.

You can use a simple display with text information and two simple buttons or a complex full color wide screen panel which shows graphics of the process in a dynamic way.

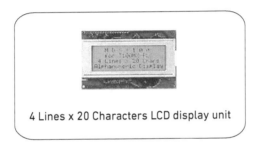

4 Lines x 20 Characters LCD display unit

Since the operator panels are very useful to translate the info from the process into understandable and well presented data, most of the automation applications will require Operator panels.

There is a dotted line that groups the PLC and the Operator Panels. This means that nowadays you can get commercial products which include both: an operator panel and a PLC as single equipment.

Not all automation projects will require an operator panel but according to the project needs you might require a way to show some information from the process. In this case you can use a simple Liquid Crystal Display (LCD).

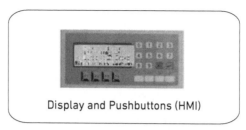

Display and Pushbuttons (HMI)

The displays are referenced by the number of characters per line and the number of lines.

For example a display could be 2 lines of sixteen characters each or a 4 lines display with 20 characters for each line, as the one shown in the former page.

There are some cases where you require entering numeric data to inform the machine or process the control settings you want. For example, a temperature set point for the control. In these cases, in addition to the display information, you need to be able to input the desired control value. For this purpose you can use an operator panel like the one shown above.

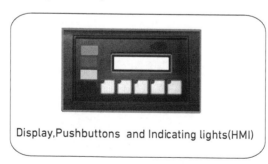

Display,Pushbuttons and Indicating lights(HMI)

When you need simplify the number of components involved, for example, when you just want to start or stop a machine, there are some other interfaces with simpler features:like the one shown above. This unit has a LCD display, three indicating lights and five pushbuttons. The program on the PLC will provide the logic to turn on or off the lights and it will detect when one of the pushbuttons on the screen has been pressed.

Technologies are succeding in smaller size designs and now you can have more complex operator panels and a very affordable pricing. For instance, you can find operator interfaces that can show full color images and not bare text.

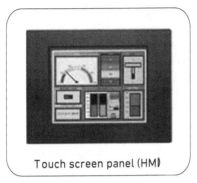

Touch screen panel (HMI)

In some cases you only need to touch the screen to press a button that has been drawn and that is included with the logic of the control program on the PLC.

Mini computers and the reduced software environment have launched a new technology in operator interfaces. They are small computer that can run more programs other than just

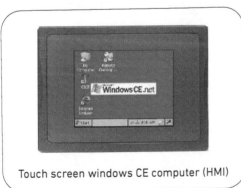

Touch screen windows CE computer (HMI)

display information. They can behave as computers with limited features which really process and handle complex processes like data bases. The Ethernet ports allow the operator interfaces to be connected to the company network and share process information in real time.

6.4   Sensors.

The PLC itself is unable to determine what is happening around. It´s supported by sensing elements or simply sensors to check for all the variables of the process.

The sensors are in charge of the detection of any status change or any new variable value. As an example, when someone opens a safety door during a machine operation, a sensor will notify the PLC about this situation. The logic in the PLC program will have to do the proper actions that can be needed to stop the entire process, sound an alarm, shut off valves, stop motors or simply notify the operator about this risky working condition.

The reliability and accuracy of a sensor are very important features, since the whole operation of a machines which depends on sensing information can be affected because of a defective sensor.

A final remark about sensors is that the sensors to be connected to PLCs influence the overall operation of the process. It's very important to differentiate from simply a sensor or from an industrial sensor. The term "Industrial" added to a product clarifies that the product can work under harsh environments or in some cases it stands multiple (millions) operations without a failure.

Of course we know that the sensor will fail after some time of operation, but the purpose is to have it operating most of the time.

Below, you will see some examples of digital sensors:

Limit switches are mechanical sensors used to detect when a machine part has reached a certain position or used a safety limits when the machine has to have a safe operation, for example, if the automation program´s designer wants to have all machine doors completely closed before allowing any operation, he will have to check for the status of the limit switches.

 Mechanical Limit Switch

The limit switch has a sensing lever that can be chosen among a lot of possibilities such as springs, wheels, wires, etc.

In order to avoid mechanical sensors that can be affected by environments full of dirt, water, grease, etc some other devices, like inductive sensors, can be used. The inductive sensor detects metal parts within very short distances (around 3 or 4 mm of distance). For non-metal parts there are the so called capacitive sensors.

Electronic Inductive Sensor

When having dangerous machines and there is a need to secure the perimeter, some sensor manufacturers utilize the " Barriers" . These ones are formed by optical (emitter and receiver) sensors. The light beam sent by the emitter can be redirected through mirrors to just one receiver. There are some cases where you can have several reception devices. This last ones send the status detection signal to the PLC.

Light Barrier

In addition to machine manufacturers, when the environment is not that harsh , the industry of home automation and security systems use very low cost devices to detect when windows and doors are closed or open. These are called Reed Relay switches. One of the devices is simply a magnet, the other is a switch that closes when the magnet is exactly in front of it.

Reed Relay Switches

To sense the speed of a rotational part, in a digital manner, most of the machine manufacturers use a device named "Encoder". There are just too many different encoder models to mention them here, but in general they can have one (named Incremental) or two outputs (named Quadrature) to provide certain number of pulses during 360 degrees of rotation.

Mechanical Encoder

6.5 Transmitters.

When using analog sensors, not all the signals coming out from them can be directly connected to the PLC input. In some cases, it´s required to change the sensed variable into a so called Standard Signal. A standard signal is a predefined electrical signal of voltage or current (for instance 0-10 volts or 4-20mA) that is equivalent to the variable being measured.

The transmitters are in charge of providing accurate information of the variable they measure and supply the electrical signal to the PLC in an safe way free of risks. The cables carrying the signal can travel long distances without any major distortion of the info.

K-Thermocouple Transmitter

Let's assume that we have to measure the Pressure of a given system. A Pressure transducer would be the device that we would use to change the physical variable into an electrical signal. The Pressure value in any specific unit, is changed by the internal electronic circuitry in the transducer into an electrical signal (could be a micro voltage which has a specific value within a range, for instance 0 to 1VDC, that corresponds to the specific pressure value.

Pressure Transducer

When the sensor does not supply the output in a standard signal, then we need to use a Transmitter. In this case the sensor is simply a Transducer, changing the variable being measured to an electrical voltage or current of any magnitude. The transmitter takes this small signal and converts it to a Standard signal(for expample 4-20mA). The transmitters are associated with the variable in order to define a name, for example:

K-type thermocouple temperature transmitter or Pt100 RTD temperature transmitter.

The transmitters are becoming very sophisticated devices and the sensor manufacturers are adding more and more features. Some recent products can read the variable, store it, show info on a local display, perform some local control if required, and finally send out the signal in the required scaling for a PLC input.

Transmitter with Display and Thermowell Case

6.6   Interface Elements.

The PLC is a control device. This means that its main function is making decisions. For such a purpose there is no need for the PLC to turn into a bulky equipment to directly manage high electrical power devices demanding control actions.

For instance, let's assume that the PLC is going to turn on a really big (in Power) motor so the current it requires, when operating, is very high. Under high voltage or current conditions, the control equipment must be isolated in such a way that the low voltage signal the PLC provides can drive other elements suitable for these demanding conditions.

Internally, some PLCs use a small relay which performs the function of isolating the control part and allowing the user to connect loads requiring certain amounts of current.

 Electronic Relays   Mechanical Relays

Most of the time, these signals are used by the PLC to notify other devices that require a simple closing of contact to start an action.When higher currents or more output contacts are required, then external relays must be used. There are a lot of relay types, with different configuration and purpose, but in general they help the PLC output to drive other higher current devices, when the operation could be risky for the electronics in the PLC.

Since the electromechanical relay is a device which has moving parts which in the time will fail. This is one of the reasons why the relay manufacturers have to specify the number of cycles the relay can make during its life. To overcome this problem there are Solid State Relays which use the semiconductor technology and don't have any moving part at all.

Solid State Relays

A device named Contactor is used to drive the high current operation of three phase motors. The control command sent by the PLC is made by sending a low voltage/ low current signal to the contactor coil.

AC Coil Contactor

6.7   Final Control Element.

These elements are devices that receive the control signal from the PLC or the interface elements and execute the desired action in terms of connection, disconnection, aperture, closure, movement, rejection or notification. Their activation reflects the final part of the control action.

AC Motor

The motor is a very common Final control element in charged producing rotation or displacement of the mechanical part to which it's coupled too. A motor can be driven directly through a contactor or through another device named Variable Frequency Drive. This component will be explained later on, since it forms part of the automation devices.

Other very popular elements are the valves. These are devices that restrict the flow of any fluid. There are manual, electrical or pneumatic driven valves. Although all types of valves are used, in automation it´s widely common to use those ones that can be activated via an electrical signal coming from the PLC.

Electrovalve

Since the operational principle is based on electricity they are called Electrovalves. They become Final control elements when they don't require any extra device to perform an action on the process. For instance, water valves that open or close for water flow are final control elements.

Electrovalves that send air to another device are not Final Control Elements.

The Pneumatic Cylinders or Pistons are very popular Final Control elements. They have two chambers that can be filled with air (pneumatic) or oil (hydraulic) to produce a displacement of the central bar. The valves direct the air or the fluid to one of the chambers.

Assume that we need to control an ejector Piston on a machine. The PLC will send the control signal to a relay, then the relay will energize the coil of an electro-valve and the valve will pass air to the Piston chamber.

Piston

From this example you can notice that the PLC is an electrical device which is the origin of the control action. On the other hand you see that the Piston is a pneumatic device that requires air to operate and constitute the Final control element.

When using industrial valves, there are situations where the torque must be somehow "amplified" in order to perform an opening or closing action. For this purpose you can find the Actuators, which are mostly located on top of the valve; based on the air direction , they open or close the big valves.

Valve with Pneumatic Actuator

6.8 Communication devices

Presently, in every automation project the communication is a must. In addition to the communication capabilities of the PLC, some extra communication devices can provide better integration. Within this list we can find:

Modems.
Wireless Modems.
GSM/GPS/GPRs.
Ethernet ports.
Serial ports.
Web servers.

Wireless Modem

We don't intend on going into details about every device since this deep information is out of scope for this book. Although it´s good to know that currently communication is the key to have a very good control system, since most of the PLCs will have to interact with a lot of devices ranging from simple sensors up to very sophisticated computer networks.

In some cases, processes are located in distant places where Internet/Ethernet connection is not available. Traditional communication devices like radio or cell phones allow wireless connection to the point where the data can be transmitted through another more popular media.

Some PLC manufacturers have designed optional modules that can be added to the PLCs in case that more communication ports are needed.

6.9 PC.

The PCs can be considered the universal tool because you need them as a part of your life. The normal life in technology is considered to be incomplete if one of two tools is missing: PCs and Internet.

In automation, a PC can be used :

1) To control a process. This is the reason why it's called PC based control. Some PC manufacturers have developed what they define as "Industrial type" PCs, but their cost, support and software tools have not evolved enough to spread the use in general industrial applications.

2) To program the PLC. The PC is used to allocate the software that simulates, compiles, edits and transfers the program to the PLC.

3) To monitor the process variables ,controlled by the PLC. For this purpose a software called SCADA (Supervisory Control and Data Acquisition) is used. This software allows a graphical representation of the process to facilitate the visualization of information and the operation of the elements involved.

4) To troubleshoot the PLC or the process. When something is wrong or there is need to use specific software to setup or adjust individual elements you can use a PC. This is why it's so important that all the elements used in automation have connectivity with PCs.

5) To store large quantities of process data. The PLC itself has limited data storage capacity which can't be compared to the hard-disks on PCs.

6) To collect all the information from different processes, PLCs or other devices and then share the information with other computer networks.

7) Providing internet access to the PLC. In some cases the PLCs don't have an Ethernet port so they are connected to

the PC through serial ports. Once the PC gets communication with the PLC, there are software tools that allow a Remote PC to access the PLC trough this serial connection.

Recently, the manufacturers of PLCs are using small computers to assemble a new control device, generating a new technology, which acts as both a PLC and a PC (with some minor limitations). The new name that has been assigned to this device is PAC.

A PAC is a Programmable Automation Controller. Some vendors simplify the explanation using the following math equation:

$$PC + PLC = PAC$$

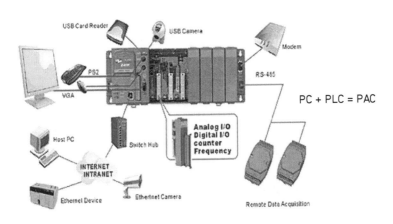

Programmable Automation Controller (PAC).

From the above graphic you can conclude that a PAC has connection ports for almost any peripheral known for the PLC or the PCs. This feature of having all type of communication and software for data processing is becoming the best reason to use a PAC instead of using just PLCs. This industry is experimenting a real growth since its debut to the automation market, five years ago.

The PAC will change the way we perform automation today since its powerful features wide the possibilities of applications, reducing the number of components involved which in time secure the reliability of the whole project.

6.10 Universal modules.

There is a brand new trend to add modules that are multi-purpose. The main idea is to make modules able to read all type of signals analog or digital.

Universal Module

Some manufacturers have launched modules to the market that can be configured to measure either voltage signals (0-1, 0-5 or 0-10V), standard current signals (4-20mA and 0-20mA) or to measure temperature through a Thermocouple sensor in any possible type (S, J, K, T,R, B and more).

The importance of these automation modules is recognized now. In many occasions, with most of PLC technologies, the automation designers are required to use different modules for different types of signals. For instance one module for voltage inputs, one module to handler current input signals and one different module to read temperature using thermocouple sensors. As a result , all final costs of the automation project. are increased. Since every channel of a single module can be set for a different variable, overall cost gets drastically reduced.

The universal modules can be connected as external modules through a pair of communication cables.

A lot of modules can be added to the same connection, defining a unique address for each of them. The cable can have long rungs and can travel all over a factory ,so several modules can be added to this network. Either near the PLC or far away from it, at the end of the cable.

If the modules need to be connected in one of the PLC slots, then it´s said that the modules have "local connection". These modules limit the size of the PLC but offer an open configuration to the user.

 Local or Slot Module

Not all the PLC manufacturers offer universal modules and not all of them are interested in making a commercial product of them, because they already have independent modules for each type of signal.

The term "universal" is applied to the programmability of the inputs but nowadays some manufacturers are conceiving that can be applied in a wider way, based on the fact that the communication protocol used (like Modbus) make them capable to be connected to almost any PLC brand in the market. Most of the PLC manufacturers have universal connectivity through this protocol. Imagine an automation application which can be easily migrated to another PLC brand by simply changing the CPU and leaving all the I/O modules.

6.11   Remote Modules.

Supported by their connectivity, the remote modules can act as an expansion of the PLC. The only difference is that these remote modules can be located on the same panel where the PLC is, hundred meters away or, as an example, in a small town of Paris... provided you have internet connection there.

There are several ways to use remote modules:

1) When the PLC is using all the modules and you need to add extra elements, then you have add a remote module on a local expansion, in the same cabinet.

2) When you have a machine of considerable length and you need to run a lot of cabling to the PLC, then you use remote modules to collect all the signals at certaing distance points and then a shielded two wire-cable will be the only requirement to connect the remote modules to the PLC.

3) When the company has a LAN (Local Area Network) or wireless LAN and you want to connect the PLC with inputs and outputs signal that can be close to a LAN port, better use a remote module with Ethernet port.

4) When a company is interested in monitoring equipment located in distant places (They could be in another city or country) a very special remote module could help. The PLC must have an Ethernet port as well.

6.12   Hubs.

When the PLC has an Ethernet port as an option module or as part of the basic architecture, the PLC for practical purposes, can be considered as a PC.

A HUB is an equipment used to provide a connection to several devices (mostly computers). Usually ,this device has many of so called TCP/IP ports that allow several devices with the same port type to share a single physical communication channel.

The electronics, in the device ,avoids to lose communication among all the other connected elements by assigning them turns, in a really fast speed.

 HUB

Since the PLC can include an Ethernet port, this is the easiest way to connect the gathered data from the process to a software application anywhere in a Local Area Network (LAN).

The HUB had been a device related more to PCs than to PLCs, but the recent technologies are forcing the manufacturers (including the PLC manufacturers) to guarantee the best communication available.

6.13   Drives

A drive is required when an automation application implies using of any type of motors and these ones require speed changes or motion control. The project can have AC or DC motors so, there are AC or DC drives also.

For AC drives it´s required to differentiate the Drive from other equipment named Soft Starter. Although the application can be similar, this last one is only used to start or stop a motor working at nominal speed (the maximum speed). In some motor´s power ranges, you must know what to do in each case, pricing is quite the same .

When you need to start or stop a motor, and the motor always runs at maximum speed (or Revolutions Per Minute RPM) you should use a Soft Starter. The connection to a PLC is limited to start/stop operation also.

The soft starter connects the motor to the power, producing a ramp on the speed until the motor reaches the maximum allowable speed. If a stop signal is applied then the motor gradually decreases the speed until the motors completely stops.

A Variable Frequency Drive or VFD is an electronic equipment that allows you to control the speed and the torque of a Tri-phase AC motor in a very efficient way. The big advantage in using a VFD to drive a motor is that the speed can be controlled at any time according to the process needs.

The drive has a special circuitry to protect the motor during operation and to change all the operational parameters such as frequency, times to increase or decrease the speed, operation modes, pre-selected speeds, maximum and minimum speeds, etc. It also has external displays and several feedback signals that can be set to the PLC or other devices for a better overall operation of the process.

The PLC can control the drive through electrical signals that can be: On and Off signals coming from contacts, voltage signals in ranges of 0-10VDC or currents of 4-20mA ranges.

If there is no PLC available the Drive can be operated through a Potentiometer or via a keypad on a small operator panel , usually located in the front.

Variable frequency Drive (VFD)

Some manufacturers are adding even more powerful capabilities to the drives, and have included a small PLC inside the drive. This enables the PLC to interact with processes and signals in a direct way, simplifying the wiring and lowering overall costs.

This is also the case for the Stepper DC Motor and Servo drives.

6.14 Voice Module.

Voice Module

This is one of the latest innovations because you can put your automation process literally to "Speak". Imagine a machine that notifies the operator about problems using pre-recorded words or sentences.

The voice module is a device that has an electronic circuit that can play pre-recorded messages. It also includes a sound amplifier in case that you want to play messages in loud noisy environments like in the case of industrial processes.

The PLC can interact with the Voice module in several ways:

- Using discrete outputs to play few messages.
- Through a binary code using several outputs to play several messages.
- Using a serial communication command to inform the module which message to play.
- Stop or play the recording according to process status.

The messages can be recorded using the multimedia software normally found on any computer.

The voice module only requires a power supply and the control signals from the PLC. The sound is like if you were listening to a CD since the voice is digitized in a very efficient way.

Questions.

1. Make a list of devices used in automation and explain the main purpose of each of them.
2. Which devices can have a PLC as part of the design?
3. How are the power supplies classified according to the internal construction?
4. What type of power supply is more suitable to absorb input voltage changes?

5. What are other names for Operator panel?
6. What is the function of a display?
7. Can you use computers instead of operator panels?
8. What is the meaning of the word "industrial" when added to the word "sensor" or any other device?
9. Explain the operation of a limit switch.
10. How can you detect non-metal parts?
11. When do you use a light barrier (optical sensor).
12. How does a Transmitter work?
13. What is the difference between a transmitter and a transducer?
14. What is a relay?
15. What types of relay you know?
16. What is a contactor?
17. How do you recognize a Final control element?
18. Explain where the name of "Electro-valve" comes from.
19. What is a valve actuator?
20. If you want to avoid wire, how do you connect equipment located very far away from each other?
21. How many different ways you have use a PC in automation?
22. What is a PAC?
23. What is and how does a Universal module work?
24. What is the advantage of using universal modules?
25. What is a remote module?
26. When can you use a remote module?
27. What is a Hub?
28. What is a drive?
29. What does RPM stand for?
30. What is the difference between a Soft starter and a Drive?
31. How can you control a PLC from another PLC?
32. What is a voice module?
33. What is the advantage of using a Voice module?

HARDWARE OF PLCS

7

This chapter studies the physical elements that form a PLC. It helps to understand the way they interact with each other in order to perform a control task.

For PLC hardware we must understand all the physical components that form a PLC system and not only the internal electronic architecture. We also want to consider the assembly parts that might be needed to develop automation projects with PLCs.

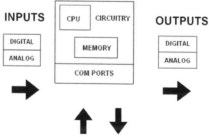

The above figure is a basic representation of the PLC parts. The arrows are showing the direction of the information processing.

You can notice that there is no other possibility for bi-directional information than the communication ports.

The PLC cannot change the status of an input or read the information of an element connected to the output.

In order to have the better understanding about the way the PLC works, let's take a look to the following graphic :

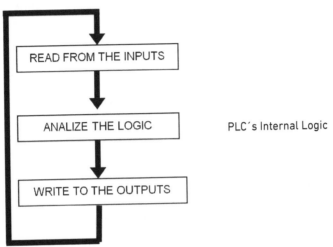

PLC´s Internal Logic

The system has to pass three main stages: Reading of all the input status, using of this info to stablish control logic and sending of the proper commands through the ouputs. The PLC keeps doing this loop consistently. It has been simplified for illustration purposes. The speed of this scanning depends on the CPU itself which is one of the main parameters to determine the real power of a PLC.

For example, if a toggle switch is hooked up, the PLC can only read if it's "On" or "Off" but it cannot force the status to turn it Off if it's currently ON.

A valve connected to the PLC output can be energized by a PLC command, but if for whatever reason it doesn't receive a voltage needed for the activation, the PLC is unable to determine if the final action, like energizing a motor. is taking place.

A feedback signal could be the solution to this problem. In this case we would use a device to sense the physical action and then send this signal to the PLC to be read through the inputs.

As a matter of fact, this is the best way a PLC controls a process: using the sensors and the logic to determine if a proper control action is taking place.

When the PLC only reads the inputs, execute the logic and activates an output then the control system is called an OPEN LOOP.

Open Loop PLC Control

On the other hand, if the PLC sends the output command and, in some way, receives a signal which doesn't produce the desired control action, then it can make quick corrections until the system is within the expected conditions. This correction process based on action- effect-result-action is called CLOSED LOOP.

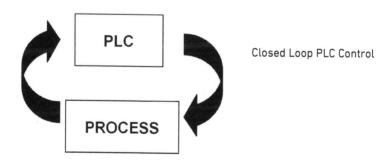

Closed Loop PLC Control

7.1 CPU (Central Process Unit)

The core of a PLC is the CPU. Similar to what happens with all the technologies available on computers, the kind of CPU used on a PLC determines the performance of the PLC. It´s related to important features such as speed, instruction and functions, communications capability, etc.

Most of the PLC manufacturers provide some basic parameters to consider when studying the processing capability. This is a the common way to compare the power of each PLC brand:

1. Contact execution time.

This parameter allows you to understand how fast the processor is, when it´s in charge of your control program. Its speed is related to the time a change, in one of the inputs, is detected.

2. Execution time for 1KB.

A control program including all kind of control logic elements, that occupies 1 kilobyte of memory size, must run as fast as possible. Since the processor is in charge of performing these tasks, it´s a very important point of comparison when analyzing other CPUs.

3. Scan time.

The Scan time is the time, taken by the processors, to read the inputs, develop the control logic and updates the outputs. It varies from program to program since the size of a program, needed for a particular automation application, depends on the application itself.

4. Maximum I/O support.

When an automation project requires a lot of inputs and outputs, the CPU must have enough capabilities to handle them all, without sacrificing the efficiency for the control program. This is one of the reasons, most of the manufacturers provide an specific maximum number of supported Inputs and outputs. This is the range where they can guarantee the normal operation of the PLC.

According to the type of CPU a PLC can behave as a normal controller or as a highly sophisticated control unit using almost all features available on a PC.

The PLC manufacturers use very specialized chips called Microprocessors and recently, more advanced units called Microcontrollers.

Every year, new, miniature, more efficient and powerful electronic devices are developed. These new technologies are generating a new world of technologies mainly in computers, which is a world where the PLCs are playing an important role. For instance, the new mini computers, like the ones found in cell phones and palm top computers are being used to develop CPUs for PLCs which are more advanced and richer in features than the Microprocessors and microcontrollers.

7.2 Memory Locations.

Also called registers, memory registers data, program and user memory.

You always need to use a memory location for several purposes. Either to store parameters that can be used later or to modify them during the process. Normally, the PLC manufacturer will tell you the range of memory locations you can use by providing the so called Memory Map.

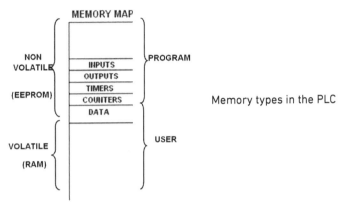

Memory types in the PLC

There are two types of memory: volatile and non-volatile .The volatile memory will be lost after you power off your PLC. This doesn´t happen with the Non-volatile memory which will remain with all the stored data. With some exceptions,the data in both memory types can be changed with normal instructions.

In some cases, through hardware, the PLC manufacturer provides an option for a backup battery so the user can install on the CPU to prevent losing information written to the memory.

This memory is completely different from the one that the CPU uses to perform its own calculations.

In this memory you can:

- Save special data.

- Store values that serve as setups.

- Store recipe values.

- Save temporary info used in program calculations.

- Use it as "bridge" or common area to share information with an external device.

The memory locations for the PLC elements are part of a total "Memory Map" which contains the way to access any of the elements of the PLC:

Inputs, Outputs, Memory, Counters and Sequencers.

The memory can be divided into two parts: the User memory and the Program memory.

The User memory is part of the total memory used to load or store temporary, and in some cases permanent, values or data.

The Program memory is used by the processor to store the control program, written by the programmer and also contains the operating system (also called Monitor Program) the processor uses for its own resources or functioning.

7.3 Inputs

Most of the information from a process is fed to the PLC through the PLC inputs.

The inputs collect all the information from the process being automated. This must be done in a very reliable way since the processor required this information to execute all the required and pre-defined logic.

The inputs can be classified in two categories:

Digital and Analog.

7.3.1 Digital Inputs.

For all the PLCs, the digital inputs are all those inputs which can be used to detect one out of two possible states:

ON or OFF status.
Item present or item not present.
Open or close.
Voltage (energized) or absence of voltage (de-energized).

All the sensing elements, connected to the inputs that can only provide one of the above outputs are considered digital signals and must be connected oe wired to digital inputs.

The Digital Inputs on a PLC can be classified as DC or AC inputs.

DC Digital Inputs:

In automation it's very common to find DC voltages ranging from 0-30VDC as safe DC values. So, values from zero up to around 1 Volt are considered as Digital "Zero". Voltages above certain limit (defined by the PLC technology) and up to 30VDC are within the range of Digital "One".

The DC digital inputs can also be classified as NPN or PNP. These terms are referred to the kind of transistor output a device can have, hence a PLC can accept.

As a general rule, a PLC with NPN inputs needs a logical Zero to activate the input. A PLC with PNP inputs requires a V+ (logical One) to determine that the input is active or energized, where V+ is the DC voltage of the power supply used.

AC Digital Inputs:

Using the same concept of having or not having a voltage to be considered as a digital signal then, an AC voltage (110 VAC, 220 VAC) applied to the inputs can be recognized by the PLC as logical "One" and the absence of this voltage as logical "Zero".

In general, there is no rule for using either DC or AC input modules. The PLC manufacturers offer them all.

PLCs with DC inputs are safer and have lower cost than the units with AC inputs, but they require a DC power supply.

Some automation developers prefer to have only AC signals in all their projects and processes, since this selection saves time and reduces wiring.

A number of external devices can be used as input devices to be connected to the physical DC inputs of the PLC:

- Safety switches.

- Start, stop, push-button, switches.

- Limit switches.

- Level switches.

7.3.2 Analog Inputs.

When the input signal can have any value (Within a specific range of values) this can be considered as an Analog Input Signal.

The analog inputs can be classified in:

Standard and Special.

a) Standard analog input signals.

These are electrical signals which have been standardized by the PLC and sensing equipment manufacturers. They can be either Voltage or Current, which can be classified within the following typical ranges:

0-1 V.
0-5 V.
0-10 V.
4-20 mA.
0-20 mA.

The analog signals are related to other automation elements in charge of changing any physcal variable into a standard signal. For instance a pressure transmitters will provide an analog signal to inform the CPU with the current pressure.

A temperature transmitter could send a signal between 0-20 mA, which is a value that represents the temperature, to the current analog input on the PLC. This particular input must be capáble of receiving 0-20mA.

b) Special purpose analog inputs:

This consideration is mainly applicable to analog input modules which can receive the signal directly from the sensor without the need of an external transmitter or any other type of electronics.

Modules of analog inputs that measure temperature through thermocouple or RTD sensors are very common nowadays.

The same case is applicable to Load Cells, where the microvolts signal provided by the load cell is connected to a module that amplifies and digitize it.

There are special purpose modules for :

- Sensors of any nature, to measure any specific physical or chemical variable.

- Weight with load cells.

- Distance.

- PH.

- Speed.

7.4 Outputs

A PLC can have several types of outputs available either in connectors or modules (Output Modules).

We are talking about the physical outputs of the PLC, (because a PLC also has internal outputs) that are used to send activation commands to the external devices such as motors, contactors, lights, valves, etc.

The outputs can be of two types as well:

Digital or Analog

7.4.1 Digital Outputs.

Similar to the digital inputs, all type of signals that can provide a true/false condition or an activation/de-activation status can be considered as digital outputs.

The Digital outputs can be:

DC, AC and Relay.

DC Digital Outputs: 0 - 24 VDC.

> The DC digital outputs can be either Sourcing (PNP type) or Sinking (NPN type). For PNP outputs, the PLC output supplies the current out to the load connected to it. In this case the power supply is derived from the PLC output.
>
> For NPN outputs, the current flows to ground, through an internal path in the PLC. The advantage of this last one is that the power supply can be external and with a higher DC voltage level.

AC Digital Outputs: 110 VAC, 220VAC.

> These outputs are meant to handle any load connected to an AC voltage source. The maximum current they can support is about 1 amp, but this current is more than enough for most of the typical control AC loads.

In case that the load requires more current but always below 10 amps, then another output type is required (like the relay type). If it's necessary to handle higher currents then there is a need to use an interface device which could be a high amperage Relay or a Contactor. This last one is very useful when having three phase loads.

Relay Digital Outputs:

The relay is a device used as an isolator between the low and high current systems. The relay is formed by two parts, the coil and the contact blocks. The relays used for these outputs are located internally in the PLC and only the relay contacts are connected to the PLC outputs.

The contacts of a relay can perform the function of a switch, which is enabled by energizing the coil through a software command to the output.

The relay output modules are manufactured with a variety of relay types with one or two contacts.

The advantage in using relays is that the outputs can be connected to any real control element, like motors, contactor, valves, lamps, etc. without any other intermediate device.

The other big advantage is because the relay output contacts can be connected to drive any loda, no matter if they are AC or DC.

7.4.2 Analog Outputs

The analog outputs of a PLC are based on a standard and can be classified in:

Voltage or Current analog outputs.

The analog outputs are capable of supplying a standard signal of voltage or current according to the specifications of the product.

1) Analog Voltage Outputs.
Typical values for standard DC voltage outputs are:

-10 to 10	V.
0-1	V.
0-5	V.
0-10	V.

Some manufactures add some other analog voltage ranges, in the order of DC millivolts, providing a very precise way to measure low voltage signals. The typical values are like the following:

0-100	mV.
0-500	mV.
-500 to 500	mV.

2) Analog Current Outputs.
Many actuators and other control elements require a current signal which has been standardized to the range of just milliamps.

Typical values for standard current outputs are:

4-20 mA.
0-20 mA.

The range of 0 to 2mA includes all the values of the other range.

To choose between either analog voltage or analog current output depends on the type of device connected to the output and the distance between the final control element and the point of connection for the analog output. For short distances it´s better to use voltage and for equipment located very far from the PLC, it´s better to use current signals. These last ones are not affected by an lectrical problem named electric noise.

7.5 Special Inputs/Outputs

In addition to the inputs and outputs some manufactures have developed special modules to better handle external signals.

High Speed inputs/counters

This type of input requires special circuitry to respond to fast changes for the input status. In many cases, this input module has its own processor to free the main CPU from such a demanding task of watching or detecting any sudden change.

Pulses at the input are counted and stored in a special memory location on the module. Every certain time (known as scan time), the locations are read by the CPU. In this case the main CPU only collects values, supplied via communications by the hi-speed module´s processor.

The high speed inputs are mainly used to be connected to pick up sensors and encoders. The first ones are just sensors or transducers which sense high speed changes and convert them in electrical pulses compatible in voltage level with the inputs of the PLC.

As explained before, encoders are devices that are mechanically attached to any rotational mechanism or part. They generate a given number of output pulses per rotation It has some special outputs which must be conneceted to the high speed inputs of the PLC module.

Temperature Modules:

This is one of the most common variables to read, monitor or control in any automation process. We could say temperature is related to any process in many different ways:

Temperature can be part of the system itself, for example when the production line uses ovens. It can be used to change the handling of the materials involved in the process by adjusting the control setups according to requirments. Sensing of temperature is useful to diagnose heat levels as a result from a faulty equipment operation.

Normally, the sensor is connected to a Transmitter which in turn changes the variable into a standard signal (for instance current). This was and still is very expensive. To overcome this situation then the PLC manufacturers designed new specialized modules to directly read the temperature from any given sensor.

Three types of sensors are widely used:

Thermocouples: Used for high temperature applications and when cost is an issue.

RTD: These are resistors that change with the temperature. Used for high accurate measurements.

Thermistor: This is a low cost sensor used mainly for sensing room temperatures.

In the past, PLC manufacturers used to offer independent modules for each type of temperature sensor, but then with new technologies in electronic circuits, new designs appeared in the market: Some modules are able to sense any type of thermocouple. In order to select which kind of sensor to use, it´s a matter of just moving a jumper .

RTD modules are more accurate and more expensive but the same options are available.

Strain gauge Module.

Since weighing systems became a very new implementation option with PLCs, the manufacturers added this type of module to their current product line. The circuitry consists of special amplifiers that receive millivolts from the load cell (the sensor) and then generate appropriate analog voltage signals that can be processed by the CPU on the module.

PWM Outputs Module.

Pulse Width Modulation is a process where the ON and OFF time of a signal is controlled in such a way that the device connected to the output will change consequently. For instance, there are proportional valves which require a

PWM signal to modulate the output. For every frequency value there is certain aperture of the valve.

The On and Off times are related on a parameter called Duty Cycle which is defined by a simple formula:

Duty Cycle= Ton/(Ton+Toff)

If the Ton and Toff times are equal then the Duty Cycle is 50%.

Communication Ports Module.

Connectivity is one of the most important needs in modern life. PLCs require several ways to connect to different devices like to other modules, to other equipment , to other PLCs, to operator panels and to programming or monitoring computers.

Normally the PLC has at least one communication Port, but when more ports are needed, some of the manufacturers offer add-on modules that have more ports of the required type: serial, USB or Ethernet.

The port operation will be analyzed next.

7.6 Ports.

One of the most important features in today's technology is the connectivity. The ability to exchange information with all sorts of devices is gaining a relevant importance.

A communication port is used to program the PLC, to connect to an external device such as touch screen panels, computers, bar code, readers, modems, etc.

Through a communication port a person, with a normal PC, can monitor/ control the execution of a PLC program or if there is a SCADA, the process itself.

For firts PLC models, only one port was included. It was used for almost everything; to program the PLC or to connect to the operator panel. But since the port only could have a function, all the connections and disconnections when a change had to be made, forced the manufacturers to find another way of programming.

New processors with more powerful hardware features were built so, new programming options arose: Instead of stopping the machine to transfer a new program the concept of ON-LINE programming appeared. Now you can make changes even with the machine running.

A simple PLC handles several communication tasks in a simultaneous way when there are more ports. For instance , a PLC could be doing all these jobs: The PLC connected to an Operator panel, registering the information from a serial temperature module, receiving the data stream from a bar code reader and being monitored by a PC with a supervisory software application.

Typical ports are.

Serial, USB and Ethernet.

Serial

RS232:

This is a very common and low cost communication port. Basically, it uses three lines to make the data exchange: TX, Rx and Ground.
The maximum allowed distance is 9 meters.

RS485:

This communication port is more reliable for longer distances. Up to one kilometer can be successfully achieved. Normally it only requires two lines (or four) to make the connection:

Tx and Rx

The shield of the cable is always connected to ground at one end.
The RS485 connection is also used for networking therefore you can connect several devices to the same two lines, but first you have to assign a different address to each of the devices connected in the network.

USB:

(Universal Serial Bus).The USB port has replaced the serial RS232   port in most of the PCs and Laptops. At the moment, just a few PLCs have evolved to include a USB port and to solve the problem, a new business has come up with the so called serial to USB converters which can work in a bidirectional way.

Ethernet:

Every industry, not to mention the whole world, is populated with computers and their respective network connectivity. Being connected to a local Area Network is a need in today's world.

Most of the industrial applications require sharing all kind of production or automation data with management software. Before, people had to manually gather the information to feed production software packages in computers. With new technologies and incorporating the Ethernet ports, all the data can be collected directly from the process without any hassle or delay.

Better developments can be expected and all the required peripherals can be similar to the ones used for computer networking: hubs, routers, wireless LAN, structured cabling, etc.

Protocols:

Talking about ports without saying a word about protocols would be like talking about people and communication without saying a word.

A protocol is the 'language" a device uses to talk to another device.

The equipment must use certain types of codes in a "logical" way for the other equipment to "understand",when sending a command or data.

When the protocol is popularly known and implemented in a wide variety of devices it's called "STANDARD".

On the other hand, when the protocol is only known by the manufacturer it's called "PROPRIETARY" and in this case, he is in the position to provide the way the protocol works or keep it only for his own products.

Some standard protocols are widely used and known:

Modbus:
Developed by Modicon in 1979, it has survived and it still is very popular today. It has two variants:

ASCII and RTU.

The first one uses standard commands for ASCII code and the second one uses a series of commands applied in so called Remote Terminal Units (RTUs).

Usually the transporting media is a RS232 or a RS485 port, but since today using of Ethernet port is very common, the protocol using this TCP/IP connection is called Modbus TCP.

Some other famous protocols are the following:

Fieldbus, Profibus, Hart, DeviceNet and Canbus.

Questions

1. Based on the PLC basic hardware graphic, which part can handle bi-directional information? Why?
2. How can the inputs and outputs be subdivided?
3. What general steps follows the CPU during a scanning cycle?
4. Explain the open loop control concept.
5. Explain the closed loop control concept.
6. Which are the most important parameters to consider when analyzing a CPU?
7. Mention three different uses for the memory.
8. Which are the types of memory?
9. What is a memory map?
10. What is the difference between User and Program memory?
11. What is a monitor program?
12. What is the meaning of "volatile" when associated to a memory?
13. How can you classify the PLC inputs?
14. Explain the difference between Digital and analog inputs. Please provide examples.
15. How the digital inputs are classified?
16. How the analog inputs are classified?
17. What are the different types of PLC digital outputs that exist?
18. Identify the possible serial communication ports a PLC could have.
19. What other Ports can be found in modern PLCs?
20. What is a protocol?
21. List some popular protocols.

SOFTWARE COMPONENTS OF PLCS

8

This chapter covers all the basic intructions that are common to most of the PLCs in the market. The chapter studies all the parameters involved in order to use every instruction.

The software instructions and functionality are the most valued parameters to consider when using a PLC, sometimes even more than the hardware. Most of the electronics in PLCs will perform an automation task and a PLC user is not fully aware of what is involved so, just in exceptional cases, hardware is a relevant factor. This can't be said from the software because it´s something any programmer will have to interact with and the acceptance depends on several factors such as: User interface, Versatility, Ease of use, Monitor program, Programming methods, Debugging tool, etc.

In general, all the PLCs in the market can do any required job, but the way of telling it how to do it, varies among the different brands. Some PLCs, can do things better than others, perhaps faster or more accurate in other cases. Most of the time the internal process execution can be transparent or not important to the user, since he only sees the final action and, considering the facts, that is what really matters.

They resemble the contact from a relay but in this case the contact is not physical but just a software representation that can be easily copied, erased, renamed, moved, pasted, documented etc. In general, all the advantages the software can offer nowadays. If a user wanted to make a change in the logic of control application, he would have to overcome a lot.

This is not the case for the software programming tool. If it has a user friendly interface, and a very easy programming technique, the design of a control program can be easier for the user.

Currently, there are several programming techniques but all of them use more or less the same type of elements:

Contacts.
Coils.
Timers.
Counters.
Sequencers.

According to the unit model or sophisticated design some other functions have been added:

Logic-Math operations.
Data manipulation.
Control Functions.

Because of the evolution of the processors in both speed and power, more complex operations (from the electronic designer's point of view) were added to simplify the the programmer's task.

For instance, a data coming out from an external device can be either multiplied or divided by a constant, compared to a setup value and the result stored on another internal data memory. All of this can be achieved using a single instruction that allows data insertion as a parameter. Without the existence of these instructions, the programmer had to take care of all the data manipulation using internal resources, provided by the PLC designer, which very often lead to control problems.

Having a more powerful set or bigger number of instructions had a relative acceptance among the programmers. Some of them consider that having such a lot of options to perform a task require a lot of time and study efforts of the instructions. before of being able to write a program. On the other hand, for a new programmer, this using of many instructions make it unclear or demanding to understand the content of the control program.

Some other programmers are interested in saving precious programming time so, the possibility of using the right instruction is very attractive. This makes the control program to look with less lines but only experienced technicians will be able to understand it.

If you are already familiar with an specific PLC brand and you don´t master the concepts quite well, the idea of changing to another brand becomes a very difficult step.

Some automation manufacturers have decided that less but powerful number of instructions are the best options for anyone who wants to start with a new brand, so they are manufacturing low cost PLC equipment with very powerful yet basic instructions set to help in the decision of using their products.

This book wants to cover the Basic instructions that, somehow, are common to most of the PLC brands. Future books written with the next levels in mind will include some remaining important instructions which are not covered here.

SOFTWARE COMPONENTS OF PLCS

8.1 Contacts

8.1 Contacts:

This element resembles the contact from an electromagnetic relay. In this case the contact is not physical but just a software representation that can be easily copied, erased, renamed, moved, pasted, documented etc. In general, all the advantages the software can offer nowadays.

The original problem with a wired system, based on physical relays ,was the fact that if a user wanted to make a change in the logic of control application, he had to overcome a lot of difficulties like rewiring, marking, adding more relays, etc. This is not the case if you use a software program.

Some contacts are very often related to the PLC inputs, but they can be associated to any other element.

If the contact is representing a device (for instance a sensor) at the moment the sensor is "ACTIVE" the contact changes its state: if it´s normally open then it closes or if it´s normally closed then it will change to open.

The contacts are mainly associated with the physical input of the PLC, but you can also have contacts from other elements such as output coils, special bits, timers or counters. These contacts can appear more than once in the whole control program.

All the software elements such as inputs, outputs, timers and counters can have contacts assigned to them.

A contact follows the same idea of a real contact found in the electromechanical relays, but in ladder logic they are simply a graphic representation.

8.1.1 Types of contacts:

In general there are two types of general purpose contacts:

Normally Open (NO)
and
Normally Closed (NC).

They refer to the current condition of the element they are associated with.

Some other contacts are considered as "special" since they add a special operational feature. Some manufacturers define their own special contacts or specify a category for them.

You can use as many open or close contacts as you want. The number is only limited by the size of the program.

8.1.1.1 Normally Open (NO) contact:

A normally open contact is a contact that remains open while the element it´s associated with, is not active or energized (in the case of a coil, timer or counter). For an input that is not active or sensing, the contact is Normally Open (NO).

A Normally Open contact is represented by two small parallel lines, like shown below:

NO contact. It´s open   NC contact. It´s closed

This NO contact is, as its name defines, normally open and only closes when the element which is representing or assigned to gets an active status.

To represent that the contact is closed, most of the software programs put a different color on top of the contact, or a color stripe over the label.

8.1.1.2 Normally Closed (NC):

A normally closed contact is a contact that remains closed while the element it´s associated with, is neither energized nor active. (In the case of a coil, timer or counter). In the case that the contact belongs to an input which is not connected nor sensing anything the contact remains closed.

A Normally Closed contact can be represented by using two small parallel lines, with an inclined line across them as shown below.

When the elements are not active, disconnected, (or not sensing anything but represented by a normally closed contact), they can be assumed as a wire that allows current flow.

Start

Stop

NC contact which is activated.
Now it´s open

NC contact which is activated.
Now it´s closed

A normally closed contact that has been activated can be recognized because of a color stripe is put on the label or directly on the contact.

8.1.2 Special contacts

Some manufacturers offer Special Contacts in their PLCs :

Immediate (Interrupt), Differential (Edge triggered) and Special bits (clocks).

8.1.2.1 Immediate contacts.

These contacts are used in cases where the input signals must be attended immediately. When an immediate contact associated to an input is activated, the normal program sequence is stopped to avoid waiting until the PLC finishes its program scanning (also referred as SCAN time). Internally, the CPU generates an "Interrupt" and goes to execute the required ladder logic for that specific immediate input. After it´s finished the CPU returns to the procedure which was executed before the interruption.

The symbol could have little variations according to the manufacturer, but the general concept is the same:

Immediate Contact

8.1.2.2 Differential or Edge triggered contacts.

The Differential contact is a feature made to reflect the transition status in the contact:

If the transition goes from Low to High, Connected to Disconnected or from "OFF" to "ON", then this is a differential contact with UP trigger or a differential contact that will work on the Positive edge.

If the transition goes from High to Low, Disconnected to Connected or from "ON" to "OFF" this is considered a differential contact with Down trigger which works on the Negative edge.

In order to better illustrate this concept, let's suppose that we have a pushbutton connected to a PLC input and internally in the program, we are using a positive edge immediate contact.

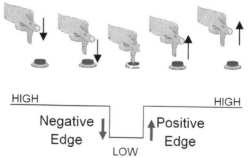

Different stages when pressing a push-button.

The input will only work at the moment we release the pushbutton. If we keep the push button pressed for as long as an hour, only one hour later, at the moment of removing the finger from the button, the logic will work.

These contacts are used in cases where the processing speed is too fast and the programmer wants to prevent the CPU from executing wrong logic. For instance, you want to make sure that the finger has been removed from a push button and not while the button is pressed.

Different symbols for Positive Edge Contacts.

Different symbols for Negative Edge Contacts.

Some manufacturers don't use the edge detection feature on a contact but on the final output which is, in general terms what we look for. For that reason, you need to use a new type of instruction like the ones shown below:

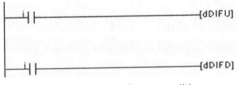

Differential Up and Down conditions

You notice that the last letter is either a "U" or a "D" . These letters stands for Differential UP or Differential Down respectively.

8.1.2.3 Special Bits

Special bits are also known as special relays, clocks and special contacts.

The special bits are independent contacts which are not associated with any input. They can close or open according to certain pre-programmed internal features.

First Scan: 1st SCAN ─┤├─

When the PLC is turned ON and the CPU finishes the first scan, the contact will close. This contact is used by several programmers to make sure that before executing a full program, certain conditions should be met first.

Always OFF contact. OFF ─┤├─
The contact is always OFF. It's used as a help of the graphical interface to separate rungs or to add a contact to the rung to avoid having the rung with no logic (some programs don't allow that).

Always ON contact. ON ─┤╫─
This contact is permanently closed an since it's not associated to any element, there is no possibility for it to change its state. You can use it to separate rungs in order to make the program more understandable.

1 Min Clock: 1 MIN ─┤├─
A 1 Minute clock is a continuous signal that opens and closes once per minute. Normally it stays ON for 30 seconds and OFF for the other 30 seconds.

1 Sec Clock: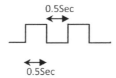

A 1 second clock is a continuous signal that opens and closes once per second. Normally it stays ON for half second and OFF for the other half second.

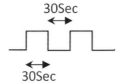

Duration of semi-cycles for 1 sec clock

The same happens for 1 min clock

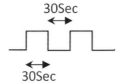

Duration of semi-cycles for 1 min clock

The PLC manufacturers offer some other Clock signals with faster speeds based on the power of the CPU.

In general terms, the special bits are used to separate rungs, to generate timer/ counters or simply as trigger of other program actions.

8.1.3 Labeling and documenting contacts:

To help you identify the contact, most of the manufacturers offer you several documentation levels so you can have either hidden or shown:

a) Name of the contact.
b) Number of assigned element.
c) The element which they belong to.
d) The wiring and marking you used.
e) The purpose of the contact.

For example:

Blue, 0037
Detects that the Doori s Open
DoorSens
Input 5
⊣⊢

Contact Information Levels.

Only the number of the assigned element is saved on the PLC memory (Input 5), the rest of the documentation is only stored in the computer file, used to develop the application.

The contacts are represented in different families of PLC as shown in the following pages.

8.1.4.1. Allen Bradley®

Contacts:

$$\dashv\vdash \quad \dashv\!\!/\!\!\vdash$$
NO   NC

Inputs

All the Allen Bradley's PLCs in old nomenclature name the inputs according to following format:

I1:i/j

Where

i is any number:0. 1, 2,etc.
j is a number ranging from 0-16 (Bit).
In the new software you can use Labels so, you don't need to worry about the nomenclature.

Example:

I1:1   Input #1.
I1:7   Input #7.
I1:8   Input #8.
I1:13 Input #13.

8.1.4.2. Automation Direct®

Contacts:

$$\dashv\vdash \quad \dashv\!\!/\!\!\vdash$$
NO   NC

Inputs

In Automation direct PLCs, the inputs are named with the following format:

X j

Where:

j is a number ranging from 0-7,10-17 (octal numbering).

Example:

X0   Input #1.
X7   Input #8.
X10   Input #9.
X14   Input #13.

8.1.4.3. General Electric GE®

Contacts:

$$\dashv\vdash \quad \dashv\!\!/\!\!\vdash$$
NO   NC

$$\dashv''\!\!\vdash \quad \dashv\!\bullet\!\vdash$$
Positive   Negative
Transition Transition

Inputs

In PLCs from GE, the inputs are named with the following format.

Ij

Where:

j is a number:0000. 0001, 0002,etc.

Example:

I0001   Input #1.
I0007   Input #7.
I0008   Input #8.
I0013   Input #13.

8.1.4.4. Siemens®

Contacts:

$$\dashv\vdash \quad \dashv/\vdash$$

NO   NC

Inputs

In Siemens PLC, the inputs are named with the following format:

Ii.j

Where

i is any number:0. 1, 2,etc.
j is a number ranging from 0-7,10-17 (octal numbering).

Example:

I0.1   Input #1.
I0.7   Input #7.
I1.10   Input #8.
I1.15   Input #13.

8.1.4.5. Triangle Research

Ldt®

Contacts: $\dashv\vdash \quad \dashv/\vdash$

NO   NC

Inputs

In these PLCs, the inputs are named with no particular format, they can use the label or name directly, prior definition in the I/O table in the Inputs table.

The table has numbering so, the Inputs have to be assigned.

first before using them.

The numbering is consecutive ranging from 0-255.

Example:

Start   in Input #1.
Stop   in Input #7.
Max_level in Input #8.
Reset in   Input #13.

SOFTWARE COMPONENTS OF PLCS

8.2 Coils

8.2 Coils

The coils are considered the outputs of a PLC. There are two types of coils:

Internal and External

The internal outputs are also referenced as marks or internal relays.

The external output coils are physical output points, which are associated with the real physical outputs of the PLC.

The internal coils are normally used to keep control of internal functions such as retentions or interlockings and, since they are only for program control purposes, they remain inside the CPU.

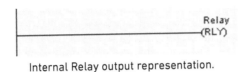

Internal Relay output representation.

The external coils will activate (energize) an external device, which requires physical wiring . They are available on the connector block of a PLC.

External Relay output representation.

If the coils are connected to the rung without any restricting element, it will be activated immediately by the PLC. If we add some logic,for instance a simple contact, the output coils will get active status only when the contact is activated.

The image below shows that only when the contact named Sensor closes, the output will be energized This same analysis can be made for the internal relay output.

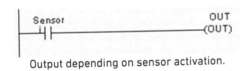

Output depending on sensor activation.

The coils are represented in different families of PLC as shown in the following pages.

8.2.1. Allen Bradley®

Coils —◯—

Internal relays

The Internal coils or internal relays in Allen Bradley are named "Marks" They use the letters B3 according to the following:

B3: i

Where

i is a number:0. 1, 2,etc.

Examples:

B3:1   Mark 1.
B3:3   Mark 3.

Outputs

The Outputs always begin with the letter "O" followed by the number "0".

O0: i

Where :

i is a number: 0. 1, 2,etc.

Some newer versions allow you to use just the letter O.

Example:

00.1 or 0:1   Output #1.
00.7 or 0:7   Output #7.

00:10 or 0:10   Output #10.
00:15 or 0:15   Output #15.

8.2.2.Automation Direct®

Coils

Internal relays

The Internal coils or internal relays are denoted with the letter "C" according to the following numbering:

Ci

Where:

i is a number ranging from 0-7,10-17 (octal numbering).

Outputs

The outputs are identified by a "Y" and use the following format:

Yi

Where

i is a number ranging from 0-7,10-17 (octal numbering).

8.2.3 General Electric GE®

Coils —◯—

Internal relays

The Internal coils or internal relays are denoted with the letter "C" according to the following numbering:

C j

Where **j** is a number: 0000. 0001, 0002, etc.

Outputs

The external coils or outputs are denoted with the letter "O" according to the following numbering.

O j

Where **j** is a number: 0000. 0001, 0002, etc.

Example:

00001	Output #1.
00007	Output #7.
00008	Output #8.
000013	Output #13.

8.2.4. Siemens®

Coils

Internal relays

The Internal coils, marks or internal relays are denoted with the letter "C" according to the following numbering:

C **j.i**

Outputs

The output coils, are denoted with the letter "O" according to the following numbering:

Oj.i

Where:

j is a number:0. 1, 2,etc.
i is a number ranging from 0-7,10-17 (octal numbering).

Example:

00.1	Output #1.
00.7	Output #7.
01.10	Output #8.
01.15	Output #13.

8.2.5. Triangle Research Lmtd®

Coils

Internal relays

The Internal coils or internal relays have their own instruction. They vary in number , according to the PLC series. There is a table for Internal Relays in the I/O table. They can have meaningful names.

Outputs

The outputs must be created first in the Output table of the I/O table. They can have a name of 8 characters long to make it more understandable. The number of outputs depends on the PLC model.

Examples:

Motor in Output #1.
Lamp in Output #7.
Alarm in Output #8.
Ejector in Output #13.

SOFTWARE COMPONENTS OF PLCS

8.3 Timers

8.3 Timers

The timers are always needed. Every process could require at least one timing operation, even for simple control actions.

Most of the PLC manufacturers show the "power" of their PLC with the number and/or the precision of the timers the programmer could have available.

Typical ladder logic for using of Timers.

Every timer has the following parameters:

a) The Set Value.
b) The Current or present value.
c) The contacts assigned to the timer.

- The timing can be triggered by any contact which should remain active (or closed) during the whole timing. According to the action to perform by the timer's output, there are three types.

- ON delay: If the contact is active, the timing action starts immediately and closes a timer contact when finished.

- OFF delay: The time only is counted just after the trigger signal goes off.

- Retentive timers: The timing operation is put on hold if the timer is somehow disabled. At the moment it's enabled again, the timing continues from its present value.

The trigger contact can be any internal or external contact associated with any element.

The timer has some topics to care about:

- Under some programming techniques, there is no need for contacts (conditions) to enable the timer.

- Every timer also has a number or name to differentiate it from the other.

- The time base is related to the precision of the timer. Usually, a multiplier number is used to define the time. For instance, if the timer is set to k50 and the time base is 0.1, the real timing will be: 50*0.1 sec = 5 sec.

- Normally the manufacturers offer two different time bases: 0.1 or 0.01 sec.

- When a timer is ON, the timing is stored in a register or memory that can be accessed by the programmer. This memory is known as present value, timer's present value or simply the timing register.

- You can use the present value of a timer to make decisions while the timing is running, like activating an output between a specific interval of time.

You will learn more about the concepts after practicing the lessons.

The Timers are represented in different families of PLC as shown in the following pages.

8.3.1 Allen Bradley®

Timers

The PLCs from Allen Bradley have three different types of timers: ON delay (TON), OFF delay (TOF) and retentive timer (RTO).

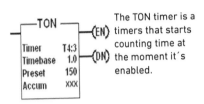

The TON timer is a timers that starts counting time at the moment it's enabled.

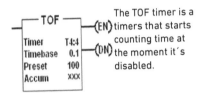

The TOF timer is a timers that starts counting time at the moment it's disabled.

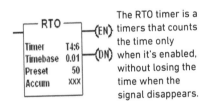

The RTO timer is a timers that counts the time only when it's enabled, without losing the time when the signal disappears.

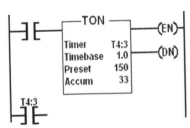

Example of connection of timer instruction.

8.3.2 Automation Direct®

Timers

The PLCs from Automation Direct have four different types of timers: TMR, TMRF, TMRA, TMRAF.

The TMRF timer is a timer that counts the time only when it's enabled. The Time base is 0.01sec. The preset value is 30 so the timing is for 0,3 seconds.

The TMRAF timer is a timer that counts the time only when it's enabled, without losing the time when is this signal disappears. When enabled again it continues from the last value. The Time base is 0.01sec. The Preset value is 30 so the timing is for 0,3 seconds.

The TMRF timer is a timer that counts the time only when it's enabled. The Time base is 0.01sec. The preset value is 30 so the timing is for 0,3 seconds.

The TMRAF timer is a timer that counts the time only when it's enabled, without losing the time when is this signal disappears. When enabled again it continues from the last value. The Time base is 0.01sec. The Preset value is 30 so the timing is for 0.3 seconds.

The Preset value in the examples on the left is a constant but can be replaced by a memory register that allows to program this value from another device like tan operator panel.

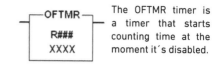

The OFTMR timer is a timer that starts counting time at the moment it's disabled.

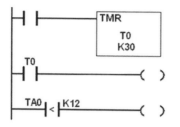

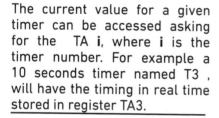

Example of connection of timer instruction.

The time base is of 1ms of accuracy.

8.3.4 Siemens®

Timers

The PLCs from Siemens have three different types of timers: TON, TONR and TOF.

The current value for a given timer can be accessed asking for the TA **i**, where **i** is the timer number. For example a 10 seconds timer named T3 , will have the timing in real time stored in register TA3.

The TON timer is a timer that starts counting time at the moment it's enabled.

8.3.3 General Electric GE®

Timers

The PLCs from GE have two types of timers: ON TIMER and OFF TIMER.

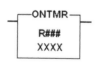

The ONTMR timer is a timer that starts counting time at the moment it's enabled.

The TONR timer is a timer that only counts time when IN is enabled. When IN disappears the timing value remains.

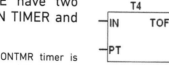

The TOF timer is a timer that starts counting time at the moment it's disabled.

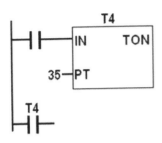

The IN input is used to enable or energize the timers. On the PT input the user can define the Preset value.

The time base depends on the timer number selected and must be consulted on the manual of the PLC used. For instance, T0 has 1ms, T1 has 10ms and T5 has 100ms of accuracy.

8.3.5. Triangle Research Lmtd®

Timers

Since Triangle Research has different PLC series, the number of timers depends on the family and can be defined in the Timers table in IO table.

The time base is 0.1 seconds for all the timers. Since the M and F series can be programmed in BASIC language, there is an easy way to access the Set value and the Present Value of any timer used in the ladder logic. It has some instructions that allow to perform data manipulation with the timing parameters.

The timer can be used as a normal ladder instruction (E10 and H series) or used inside a Customer function which is written in Basic language (M and F series).

SetTimerSV: To program the SET VALUE of a given timer at any time.

GetTimerSV: To ask for the SET VALUE of a given timer at any time of the program execution.

TimerPV[n]: To ask for the PRESENT VALUE of a given timer at any time.

DELAY: To generate a small delay when running a small program in Basic in a Customer function.

Examples:

SetTimerSV 1,2000.
Sets the timer 1 for 200 sec.
SetTimerSV 2,
GetTimerSV(1)+10.

Sets the timer 2 with the value of another timer (the number 1) and add the value of 10 (1sec).

SOFTWARE COMPONENTS OF PLCS

8.4 Counters

Counters.

Like the timers, the counters represent a very important feature to determine the final capabilities of a PLC.

Counters are used to count a lot of possible sources, such as machine cycles, produced pieces, stop times, etc. In many occasions they can count pulses from a Encoder to determine a position.

The counter has the following parameters:

Typical ladder logic for using of Counters.

a) The Number or name of the counter itself.
b) The Set Value of the Counter.
c) The Current or Present Value register of the counter.
d) The Reset signal for the counter.
e) The contacts assigned to the counter.

The counters can be classified as:

Incremental, Decremental and Reversible.

The incremental counters can count from zero up to a certain user defined value.

The decremental counters count from a given value down to zero.

The reversible ones have two inputs one to count up and the other to count down.

When a counter reaches its maximum value and the corresponding input changes then the counter closes its contacts, resets itself and starts over.

Some manufactures don't have a RESET input for the counter, but they provide an instruction instead.

High Speed Counters.

When the counting inputs have a high speed, a normal counter can work incorrectly. This is because the PLC uses a time to execute all the control instructions in the program and some milliseconds will pass before executing the instruction of reading the status of the counting input. To solve this, a conjunction of High Speed inputs, interrupts and special instructions are used to avoid missing counts.

For critical applications, where the inputs must react at higher speeds, software counters can't be used and modules with specific hardware must be required instead.

These modules have a special circuitry with stand-alone processors that don't occupy processing time of the controller´s CPU, holding the data until it.

The Counters are represented in different families of PLC as shown in the following pages.

8.4.1. Allen Bradley®

Counters

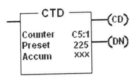

In some PLCs like the Micrologix there are 256 counters which can be named from C0 to C255. Once you have selected a counter you need to use a different number for any other type of counter you plan to use in the rest of your program.

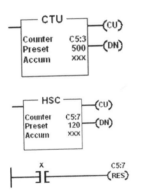

X can be any of the following:

1. A NO contact from an Input, to reset the counter manually.

2. A bit corresponding to the counter number, like the following ones:

C5:#/13 Done. This bit's on when the counter reaches the preset value. You can also use the following notation C5:#/DN.

Some other bits can useful when working with counters, here you will find some of them:

C5:#/10. Update accumulator bit. This bit's on when the accumulator of the counter has been updated. (Also C5:#/UA).

C5:#/11. Underflow bit. This bit's on while the counter's accumulator is below the preset value. (Also C5:#/UN).

C5:#/12. Overflow bit. This bit's on while the counter's accumulator is higher than the preset value. (Also C5:#/OV).

8.4.2 Automation Direct®

Counters

Counters in Automation Direct PLCs are define as CTX. Where X is a number with octal numbering (0-7, 10-17, 20-27, etc). Once you have selected a counter you need to use a different number for any other type of counter you plan to use in the rest of your program. There are two types of counters. The normal ones and the up/down counter.

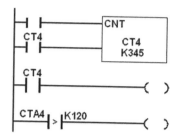

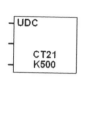

Example of how to use a counter.

In this case a contact from the same counter is used to Reset the counter and then to activate an output in the next rung.
You can ask for the counter value at any time and use this information to make decisions as in the last rung.

Memory registers:

The memory registers are denoted with the letter V and , depending from the PLC range from V1400 to V1777 in octal. If you use a memory position for any of the elements of the PLC, somewhere in the program you need to use some instruction to store a value on that specific memory register.

The counter CT4 counts from 0 up to the Preset value, which in this case is 345. The counter has two inputs, the upper one is the pulse input, the lower one is to Reset the counter.

The UPDOWN counter CT21 is a counter that starts counting Up if it receives pulses in the upper input. It counts down if receives pulses in the middle input. The lower input is the reset input.

8.4.3 General Electric GE®

Counters

Counters which can be named from C0 to C255. Once you have selected a counter you need to use a different number for any other type of counter you plan to use in the rest of your program.

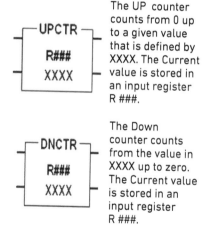

The UP counter counts from 0 up to a given value that is defined by XXXX. The Current value is stored in an input register R ###.

The Down counter counts from the value in XXXX up to zero. The Current value is stored in an input register R ###.

XXX can be a register as well. The Registers can be selected from R1 up to R500.

8.4.4. Siemens®

Counters

In some PLCs like the S7-200 there are 256 counters which can be named from C0 to C255. Once you have selected a counter you need to use a different number for any other type of counter you plan to use in the rest of your program.

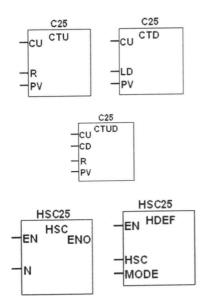

Depending on the CPU you can have up to 6 high speed counters which range from HSC0 up to HSC5.

8.4.5. Triangle Research®

Counters

The counters are created in the Counters table in the IO tables. E10 and H series have less counters than more sophisticated series like M and F.

The Counter can be used as a normal ladder instruction (E10 and H series) or used inside a Customer function which is written in Basic language (M and F series).

SetCtrSV: To program the SET VALUE of a given counter at any time.

GetCtrSV( ). To ask for the SET VALUE of a given counter at any time of the program execution.

CtrPV[n]. To ask for the PRESENT VALUE of a given counter at any time.

Examples:

SetCtrSV 3,2489.
Sets the Counter 3 to count up to 2489.

SetCtrSV 4, GetCtrSV(1)+15.
Sets the Counter 4 with the value of another counter (the number 1) and add the value of 15.

SOFTWARE COMPONENTS OF PLCS

8.5 Sequencers.

Sequencers.

In some cases you need to develop control programs where there is no need for a complex logic other than having the outputs activated in a certain sequence. This can be easily understood if you have observed the way the music boxes work.

A music box is indeed a sequencer. It´s formed by a drum with a lot of pins that can be spaced all over the area of the cylinder as shown in the below picture.

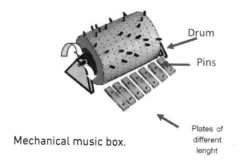

Drum

Pins

Mechanical music box.

Plates of different lenght

Next to the drum there are a certain vibrating plates of different length that can be reached by the pins. Being of different length means that they can vibrate at different frequencies making each of them produce a different sound.

The melody is created by the combination of sounds and in this case by the combination of pins touching a specific plate.

You will notice that at a given time you can have a pin touching one, none or all of the plates. This is your decision: the way you combine the pins to produce a melody.

The drum can rotate at a fixed speed to guarantee that the melody is following a rhythm.

The sequencer is the drum, the rows of pins that touch the plates can be assumed as the step selection and finally the plates can be assumed as the PLC outputs.

For instance, let's assume that we want to activate the outputs one by one until we reach the last one, then we must start the cycle again from the first output to the last one. You could think that this is a sequence that follows the following steps:

> 1st Step... Turn On Output 1.
> 2nd Step... Turn Off Output1 and Turn On Output2.
> 3rd Step....Turn Off Output2 and Turn On Output3.
> 4th Step... Turn Off Output 3 and Turn On Output4.
> 5th Step... Turn Off Output4 and Return to 1st Step.

This would be a sequencer of 4 steps since the 5th step is the same as the 1st step.

If we use a Lamp for every output, you could see the visual effect of the sequence as the one shown below:

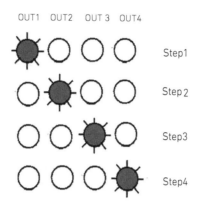

The speed of jumping from one step to the next one is determined by an internal signal that, according to the implementation and the PLC technology, can be one of the following:

- A special bit or clock signal.

- A contact signal coming from a sensor.

- A contact signal from a timer or a counter.

In some powerful sequencers, for every step you can select the type of signal that will produce the jump to the next step.

In order to specify a sequencer, you have to consider the following concepts:The Sequencer number or name.

- The Sequencer number or name.

- The number of steps.

- The jump to a specific sequencer's Step.

- The Reset of the sequencer.

8.5.1 Sequencer Number or Name.

Since a sequencer requires a lot of internal resources from the CPU, the number of Sequencers of a given PLC is always limited. If the PLC processor is very powerful a good number could be from 5 to 10 sequencers. Most of the manufacturers of big PLCs allow you to have more than one sequencer and in order for you to know the sequencer you are using, you are allowed to put either a number or a label, as you can see on the graphic shown below:

Sequencer coil representation.

The sequencer can be considered a separate instruction. In the example the instruction is "AVseq". The name of the sequencer is "Seq1" but could have a different name like "SpecialSeq".

8.5.2 Number of Steps of a Sequencer

The number of Steps, as the number of sequencers, on a PLC is also limited.

The maximum number of steps varies from manufacturer to manufacturer, but usually it´s around 30 steps.

Every step can have a contact like the one shown in the following figure:

Seq1:3
—| |—

Contact for Step 3 of Sequencer 1.

The notation is as follows SeqX:Y where X can be replaced by the sequencer number (if its allowed) and Y is the Sequencer Step. In this case we are talking about the third step of the Sequencer 1.

8.5.3 Jump to a Specific Step

In some occasions you do not want to continue the normal sequence but want to jump to another step in the sequence. It could be before or after the current step the sequence is at. These jumps must be always within the same sequencer. For these cases you need to use a special instruction which, after fulfilling some logic, will allow you to jump to the desired step.

ManualRes   Seq1:2
—| |—————————————————————[StepN]

Jump to step instruction.

In the above graphic, we are using a ManualRes contact that can be a pushbutton to jump to the Step 2. This means that no matter the current step the sequencer is on, if we press ManualRes pushbutton, the sequence will continue from Step2.

8.5.4 Re-starting the Sequencer

Since you don't need to use the maximum number of steps, but only the ones really needed for your control action, you need to have a way to re-start the sequence at any time. This can be done by using the Reset sequence instruction. Most of the sequencers use a counter to count the steps, so this signal is equal to the ones used in counters to reset the counter.

Normally, the logic used to reset the sequencer is a step number (to have automatic re-start) or another external input signal like a pushbutton in order to re-start the sequencer in a manual form at any time the user wants.

Instruction to Reset the Sequencer.

The graphic above shows that the Step number 5 will reset sequencer one.

To recapitulate, a complete program using a sequencer would look like the one shown below

Notice that at every step we can activate the outputs that we want. In the second step outputs 2 and 4 are activated at the same time.

Some manufacturers offer you the possibility of having a sequencer like a single graphic instruction. It´s like having a table where you can make your selection of all the sequencer functions: how and when to jump, what to activate at any given step, when to re-start the sequence, etc.

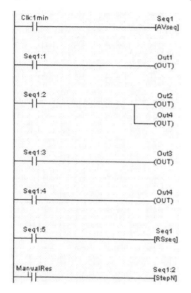

Complete sequencing program

8.6 Logic and Math Operations

Though these instructions are not the main focus of this book, since they are going to be covered by some of the future books, it´s convenient to have a brief explanation.

The information inside the PLC can come from different sources: the current time from a timer, the set value of a counter, the current step of a sequencer, the speed of a motor being controlled by the PLC, etc. All this information is handled as a group of zeroes and ones (0 and 1 are the numbers of a Binary code) within the registers and memory of the PLC.

For some PLCs, it´s possible to make use of this data through a set of new available instructions.

To perform decision and calculations you need to use operators:

Decision: <,>,=,≠, ≥.
Math: ADD,SUB, MUL, DIV,MOD.
Logical: AND, OR, XOR, EXOR,SHIFT, ROT.

For instance you could say:
If the temperature is higher than 300 °C then output 3 must be active. In ladder logical it could be:

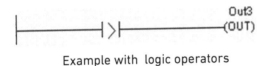

Example with logic operators

The logic symbol has one data at each side: temperature and K30. Notice that the contact has an internal symbol of "Greater Than", meaning that the contact will compare the two data.

In this case, the variable Temperature is calculated somewhere in the PLC program and stored on a memory, register or variable. The constant 30 corresponding to the 30 Celsius degrees and it´s accompanied by the letter K, to state that it´s a constant different from a character code.

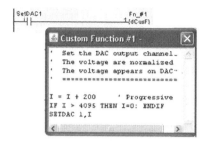

One input signal will execute a Basic language program.

Most of the PLCs have instructions to represent any of the math operations, however if you need to develop complex calculations it becomes a difficult task. For that purpose some other PLC vendors have implemented high level languages like C or Basic. It allows the user with a very complex calculation to write a program using standard programming instructions.

The mix of ladder logic and a programming language is a powerful and rich combination.

In the picture, if the input is active a new program written in Basic language, will be executed.

Functions: The more software capabilities a PLC can offer the faster or easier the user can write a program.

Some manufactures offer one or two types of functions:

1. Pre-defined functions to perform repetitive instructions such as moving a table of data from one memory range to another.

2. User-definable functions where the programmer can write his own formulas and programs and execute them as a program subroutine which can be called at any time from other parts of the main program.

This last option is very important since it will save a lot of time when programming PLCs.

Again, you don't need to worry about these instructions since you will learn and practice them on our second book.

Questions

1. Which are the software elements used to write a ladder logic program?
2. Discuss what you think is better: Simple basic instructions or a large number of powerful instructions?
3. How can a NC contact be represented?
4. When do you know that a NO contact is closed?
5. Where do the contacts in a ladder program come from?
6. In the software, how is a contact represented when it has changed to the opposite state?
7. What is an immediate contact?
8. What is an edge operated contact?
9. When do you have positive and negative edge detection?
10. What are the so called Special Relays?
11. What is the purpose of using a first scan contact?
12. Why is the operation of the special bit contact named "Always ON"?
13. Explain why the Coils are Internal or External.
14. Can a coil appear more than once in a ladder program?
15. When do we say that a coil is energized or activated?
16. Explain the operation of all the types of timers available in the ladder logic.

17. For a timer, what is the Time base?
18. What is the present value on a timer?
19. What happens with the timing, if through the ladder logic, we de-energize the timer at any time?
20. How many contacts can we use from any given timer?
21. How do you know the timer accuracy?
22. What is the minimum time we can achieve with a timer?
23. Explain the parameters to consider when using a counter.
24. What happens if after some counts, a counter does not receive more pulses?
25. What happens if the counter RESET is always ON?
26. What is an Incremental Counter (Up counter)?
27. How does a Down counter work?
28. What is a reversible counter? How does it work?
29. When do you use a software high speed counter or an external high speed counter module?
30. Which parameters do you have to consider when using a Sequencer.
31. When would you prefer to use a sequencer instead of rungs with pure ladder logic?

LADDER LOGIC PROGRAMMING

9

This chapter is a comprehensive study of ladder logic programming and introduces the rules to complete a successful PLC control program.

The use of ladder logic is very extended and even with the existence of a lot of new programming methodologies, the ladder logic is still very popular. Among other reasons, because it´s really related to persons with electrical background and those are the ones who were in charge of performing automation and PLC programming. This asseveration is no longer valid since some other disciplines of human knowledge have learned to develop control programs using programming methodologies close to their own interests.

Presently, we could say that ladder logic programming is applicable to a hundred percent of the projects. However we need to clarify that there exist situations where another methodology produce better results in programming time, easiness, troubleshooting or simplicity.

9.1 Changing an Electrical Diagram into a Ladder Circuit.

To better understand the concepts of a ladder program, let's first consider four physical elements: A light bulb, a switch, an AC outlet and the male plug.

From the graphic, you can notice the way we have interconnected the different devices in order to have an operative electrical circuit.

For practical reasons we have omitted using a fuse as a safety device against shortcircuits.

One terminal of each device is connected to one of the terminals of the other. The AC outlet is the power supply used to supply energy to the lamp, once we close the switch.

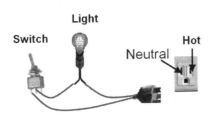

Electrical diagram to control a lamp.

The electrical representation for the above circuit is shown on the next page.

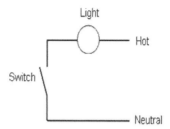

Real electric circuit to turn a lamp On or Off

Every device has been replaced by a symbol that allows you to have an easy understanding of the whole operation.

While the switch is open, there is no current flowing through the circuit. To turn on the lamp we need every terminal of the lamp to receive a terminal from the power supply.

When you close the switch, this device can be considered as a normal wire, and then considered as a device that is not present, allowing the current to flow without restrictions.

On this particular case we have an AC source which generates a current that flows first in one direction (from Hot to Neutral) and then the opposite direction (from neutral to ground), sixty times per second (also known as frequency).

If we replace the power supply and use a DC source instead, the current will flow in one direction only.

The electrical representation can be redrawn. Notice that the power supply terminals Hot and Neutral have been drawn as two vertical lines that can be prolonged downwards when more parallel circuits are needed.

In general you can assume that the two vertical lines are the power supply, no matter if they are from an AC or DC source. All the circuit elements, can be drawn in a horizontal way.

Re-drawing of the electrical diagram for the lamp

The ladder representation uses the same analogy as the electrical diagram. You can see however that the vertical terminal of the power supply has been removed. Every electric symbol is replaced now by a graphic representation that must follow certain rules.

Ladder logic program to control the lamp status

9.2 A ladder Program.

A ladder program is based on the combination of the software elements you already know and a lot more found in other advanced PLCs, in order to develop a control system.

The ladder programs are formed by rungs, which can be as

many as the PLC's program memory allows. In general a RUNG is formed by Logic and Actions.

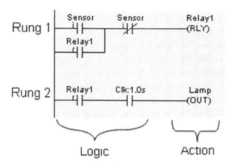

Rung, logic and action concepts.

Just below the last rung, some manufacturers use an END action to notify the PLC program the scanning limit.

On each rung, a combination of contacts (logic), determines the activation of an output (action).

Consider the following ladder program:

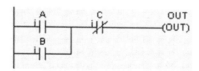

Ladder logic to activate an output

On the first rung, at least one of the electrical paths must have all the contacts closed to activate the Output relay. From the graph you can see that there are two possible paths.

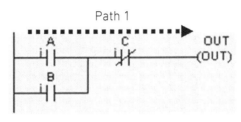

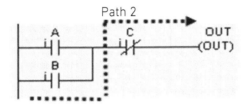

Two different paths to activate an output

The output will remain "connected" as long as the contacts are closed.

For the logic you can have contacts, connected in series or parallel, coming from:

Inputs, outputs, relays, timers and counters. Less frequent the contacts from sequencers and some logical or math instructions.

As for the actions you only can have: Outputs, internal relays, counters, timers, level bit logic, math operations and functions instructions to handle data.

9.3 Master Control Relay.

One of the most common applications on electric circuits, that is still applied when most of the so called wired logic is being replaced by software logic, is the Master Control Relay.

There are two ways a Master Control Relay (MCR) is used: implemented with a real relay or through a software relay.

The real relay is an external relay with several contacts. The relay coils is energized through some safety physical elements like push buttons, emergency stops and safety switches.

Since the intention of the MCR is to prevent the control operation and activation of outputs under an emergency situation from improper or unsafe operation, the different relay contacts are used to block the power to the output modules.

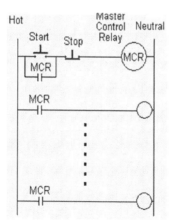

Usage of Master Control Relay on a Ladder logic Program.

The PLC itself can remain energized and active, but since the voltage has been removed from the PLC outputs, all the final control elements connected to the PLC, are forced to the safe condition which is the status the should have when the whole system is off.

The reason to keep the PLC energized is, once the emergency is resolved, that the CPU processor can know what it was doing before the sudden stop and then contuing from there.

If required, one of the MCR´s contacts can be used to notify the PLC through its inputs that the MCR has been deactivated and that it should perform specific and proper control actions related with the emergency stop.

On the other hand, the software master relay control is a software element, available in some PLCs, that allows you to disable a group of rungs, located between the instructions. Its function is similar to the physical relays, used to disconnect a circuit.

When the CPU finds that there is a MCR active, all the outputs in subsequent rungs are turned off. The process is repeated every scan until the MRC becomes active again.

Some PLC manufacturers have defined not to use a MCR instruction itself, since it can be implemented using other common software elements.

Carefully observe the ladder program on the next page. You can see that CO appears several times all over the program. This means that in order to allow the operation of the rung where it appears the output CO (our MCR) must be active.

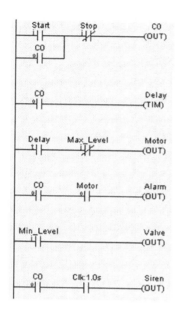

Replacing the MCR with normal
PLC instructions.

Regardless the function of any control program, all the contacts that belong to the same element and are connected alone with nothing in parallel (like the contacts from C0 or the contacts from any MCR, are known as "interlock" contacts.

The interlocking is a way to "authorize" the operation of the rung, since if the contact is open, no matter the status of the remaining elements, the rung will not operate.

The program shown on the next page has the same operation as before, but it introduces two new instructions, which are particular to a PLC manufacturer, named ILock and ILoff that clearly isolates parts of a given program.

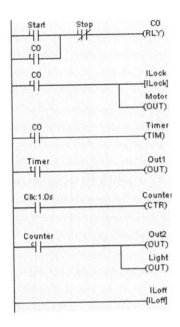

Using of interlocking instructions.

Exceptions are made on the first rung, the rest of the program is between the ILock and ILoff instruction, meaning that this group of rungs in between will only work when the Interlocking instruction is active which, in time, is depending on C0 .

The program will run only when C0 is energized, otherwise it's stopped because the interlocking instructions are not active. When the interlocking instructions are used, the scan time is drastically reduced since the CPU doesn't check for the logic involved and practically ignores this part of the program.

9.4 Typical ladder programs.

There are several typical configurations you normally use for a ladder program.

9.4.1Start/Stop.

You can use these rungs in most of the programs you write. It will give you a way to Start or Stop the program execution. This is also called the MCR, but implemented inside the PLC.

It uses a NO pushbutton named START and a NO pushbutton named Stop. In case that any of the physical pushbuttons is NC, then you need to change the program by using the opposite symbol: a NC contact for Start or NO contact for Stop.

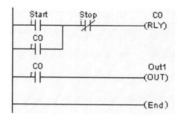

Program to control a given output

Observe that the last rung is an END command. This has been put for illustration purposes only. As mentioned before, some programming software will require you to put it.

This last rung must be at the end of the whole program and it's only a method used by the CPU to mark that the program must be stored and executed up to there.

The Start/Stop can use an internal or external output. In this case we are using an internal relay output named C0. Only one contact is used now, but this is because the program is of rungs only. You can use as many contacts as you want in other parts or long programs.

9.4.2 Using timers and their contacts.

Using of timers in ladder logic programs is very common. At any moment you will need a delay or to wait some time before starting a new action.

You only need to define the way to activate a timer to use it. The program shows a contact from C0 that is used to enable the timer.

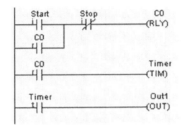

Simple program to use a timer.

If for any reason you don't use any logic to enable the timer, and the timer has a direct connection to the bus bar on the left, the timer will start counting time as soon the PLC is energized.

Using a timer implies the use of the information it provides: the total timing or the present value which is the time counting value it has, at any given time, during the timing process.

The elapsed time is counted from the moment of closing of the enabling contact associated to the timer.

The timer's present value is a memory register that can be used (by comparison) to trigger and action even when the timer is still counting time.

You can use as many timer´s contacts as you want and locate them on different rungs of the program to obtain specific enable signals.

The rungs containing timer contacts can be located before or after the rung where the timer is used.

9.4.3 Using counters and their contacts.

The counters are elements that also can work with or without an enabling signal. However they require a signal that must open and close to activate the counting action. A counter doesn't count by itself, it requires input pulses to increment or the decrement the counter value.

The program below shows that the enabling signal is C0 and as mentioned before, can be omitted from the ladder rung, conce the Coil C0 is energized.

On this case, the pulses are automatically generated by an internal "clock" where the contact continuously opens and closes once per second. Since the counter is fed with these pulses, the action performed by this particular counter is to count seconds.

Simple program to count objects.

The Reset signal has been omitted on purpose. But always keep in mind that if the counter has no way to reset, then it won't normally work as it should but only at the initial conditions, when the power is applied to the PLC.

In some cases, the symbols used for counters force the programmer to make all the proper connections before using it. All the inputs of the counters must be connected.

In other cases, like this one, where the reset signal is separated from the counter, sometimes it´s required to reset the counter at least one rung before counter's rung. This is because of the scan of the program and the nature of the up or down counting.

Somewhere on the programming software you need to specify, the value the counter is going to use as a setting. For example, if you plan to count up to 34, you must specify this number. To do this, most of the PLC manufacturers use a memory location which is not visible in the ladder representation. Some others allow you see it just on top of the symbol.

The information that can be obtained from a counter is the value to count and the current value of the counter at any given time during the counting process.

The value to count is set by the closing of the contact associated to the counter.

The counter´s present value is a memory register that can be compared with given values to provide an output without the need to wait for the counter to reach the final counting value.

You can use as many counter contacts as you want and locate them anywhere in the rungs of the program to obtain specific actions that depend on the counter.

9.4.4 Using the current values of timers and counters

If the PLC has some comparison instructions it's possible to improve the performance of the ladder program by using all the current values of changing registers like the current values of timers and counters.

Program to perform control actions at specific or given
timing and counting values.

Examining the ladder logic program shown above you can see that in the rungs 3 and 6 there are some new instructions, which have not beeb used or explained before.

The instructions are called "Comparators" and they take two values or registers and according to the instruction type, the associated contact will close at the moment the comparison is valid.

For instance, the instruction in rung 3 is comparing the current value of the timer with the number 15. While the current value of the timer is Less than 15, then the instruction is valid so the contact is closed. When the current value overpasses 15, then the contact opens since the instruction is not valid anymore.

Analyzing the control rung and assuming the timer is counting up to 50 (no matter the units) we could say that the output "Out2" is going to be ON only when the counting value (or the timer current value) is below 15. The rest of the time the output is going to be off.

Some PLC manufacturers identify the current value of the timer by adding another letter or using a different name in order to differentiate it from the name of the timer contact. For instance, the timer contact is called simply timer and we should use another name for the current value like TimerCV.

The same analysis can be made for the Counter, like the ones on the rest of the rungs.

The comparison contacts can be: Equal, Less than, Greater than and Equal or Greater than.

Every comparator requires two operands that can be either two registers, a register and a fixed value or constant , and a register with a memory location.

Instead of using several timers and counters, you can use logic comparators to enable control logic that only depends on one timer or one counter. Only if the control program allows it and the PLC supports these instructions.

9.4.5 Combining Timers and Counters.

It's very common for a program to have timers and counters

at any part of the ladder program.

You can use a timer to produce a delay for certain action, to enable a rung or simply as prevention of activation.

For instance, in the below program, once initiated by pressing Start, will wait for some time before allowing the counter to count.

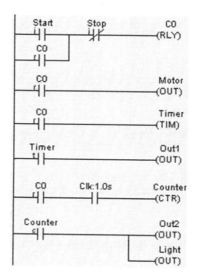

Using timing and counting actions in the same program

In other words, the counter won't start counting unless the timing action has finished and its contact has closed.

Again, this program is for illustration purposes and we have omitted the reset rung for the counter.

9.4.6 You can use contacts from activated devices or elements

There is not much difference in using an enabling contact or a contact from the resulted action.

If the contact is enabling, let's say an output, we can use the contact from the output as equivalent enabling signal to the first original contact that closed. To clarify this concept, let's take a look at the ladder program below.

On Rung 2, a contact from C0 is activating an output named Motor.

Later on the Rung 3 the same contact from C0 is activating a Timer. It's exactly the same to use a contact from the output Motor instead. This is because the time difference between the activation of C0 and Motor is just milliseconds and it wouldn't be noticeable for the human eye, nor for the operation of the program. So, in practical terms we could consider both activations as occurring "simultaneously".

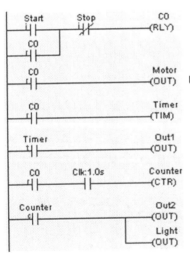

Using the same contact in different parts of the program.

Whether you are using contacts from C0 or Motor, it's up to you, the programmer, to have a clearer logic or to make the program more understandable for troubleshooting.

In some cases, with the contact C0 it´s easy to understand that it depends on having the machine ON. In other cases, when a contact from Motor is enabling the timer it´s very easy to understand that the Motor must be ON before the timer can start counting any time.

On the ladder diagram shown below, you will notice that the last rung is using a contact from Out3, and this Output is activated by the rung located just before. We could eliminate the last rung and add the output of Out4 just below the output Out3. In this case all the outputs Out2, Out3 and Out4 would be driven by the contact from the Counter.

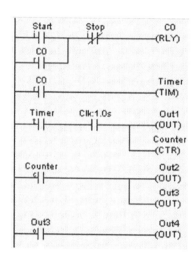

To activate Out4, it´s the same to use the contact from the counter (Counter) or the output (Out3) which is previously activated by it.

9.4.7 Using sequencers.

The sequencers, as mentioned before, are very useful when the activation of outputs is following certain "pattern" or done in steps that are triggered by an input or an internal clock. In those cases, it's much better to use a sequencer rather than a ladder program. In fact the sequencer replaces a lot of ladder logic rungs.

To use sequencers in a control program you should use the following configuration:

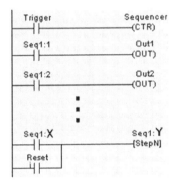

Typical program for using of
Sequencers.

The trigger signal can be any contact from any of the other remaining elements Input, Output, Timer, Counter or Internal bit.

The letter X represents the maximum step the sequencer is going to reach (in this case is the Set value of the counter).

The letter Y is the step value we want the sequencer to re-start. Note that we have added an optional Reset input which will help us to reinitiate the sequencer at any moment we want. The sequencer uses a counter.

As a general rule, you can combine all the step contacts to activate an output. For example we could add a parallel contact from Seq1:7 to the one we already have noted as Seq1:1. This is shown on the next program.

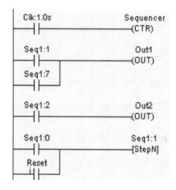

Sequencer for two outputs.

According to the combinations of contacts and the number of sequencer steps you can develop a lot of effects on the outputs.

9.4.8 Edge triggered events.

The use of this feature varies from manufacturer to manufacturer. This edge feature can be applied to the edge detection of the contacts or to the outputs.

The model to apply a differential up or down transition is as follows:

For Differential UP:

```
    | Sensor                    Out1
    |--| |----------------------(dDIFU)
    |
```

Differential up or positive edge output activation.

The output will be energized in the precise moment when the Sensor is detecting, at the moment its contact is closed.

For Differential DOWN:

```
    | Sensor                    Out1
    |--| |----------------------(dDIFD)
    |
```

Differential down or negative edge output activation

If you use the other output function of Differential down, the output will energize only at the moment the sensor changes its condition from detection to non-detection.

For both cases the actions only occur during one scan time cycle so, they are not really visible.

Questions.

1. Define the term "Rung".
2. Define the term "Logic".
3. Define the term "Action".
4. Which elements can appear on the Logic part of a rung?
5. Which elements can appear on the Action part of a rung?
6. For a ladder program, what is an electrical path?
7. Explain why a Master Control Relay is important.
8. What is interlocking?
9. Which elements can you use for a start/stop circuit?
10. Can you implement sequencers through ladder logic?

Notes

OTHER PROGRAMMING METHODOLOGIES

10

Ladder Logic is not the only methodology to program PLCs. This chapter explores basics of some other programming methods.

Since every manufacturer has developed his own particular way to program his equipment, every time you as a user wanted to change the current PLC brand you needed to rebuild the whole program for the new brand. The first standard is the IEC-1131 and was developed in Europe. During recent years this standard has had variations and turned into IEC61131-3.

The idea of having a common programming technique has offered six possible ways:

> Instruction List.
> Structured Text.
> Ladder Diagram.
> Function Block Diagram.
> Sequential Flow Chart.
> Flow Diagrams.

10.1. Instruction List (IL).

Also known as mnemonics. This method uses commands which are written just with three letters. Though it's a cumbersome way to work with, the processing speed is better than the other techniques.

        STA (* Save variable*)
        LD
        MOV
        OUT

This is the same programming method used in assembler language which is very popular with microcontrollers.

To add comments you only need to put the expression between asteriscs.   **(* and *).**

10.2. Structured Text.

You use the same methodology as writing programs in high level languages like BASIC or C++. There are commands, created just with the purpose to perform control such as activation or deactivation of timers, set or reset of internal or external coils. In general, a complete set of software commands allow the user to replace a part or all of the ladder instructions.

The programmer can use a lot of powerful instructions to develop a control algorithm:

        IF...THEN..ELSE
        CASE
        FOR .... TO
        WHILE....

REPEAT...

Every program can be saved and used in other applications. calling it as a procedure or subroutine.

10.3 Ladder diagram (LD).

This is the most popular method to program PLCs, because most of the PLC manufacturers decided to implement it as standard.

It follows the electrical rules of connections so it's an electrical programming type.

This method is covered on the other chapters.

10.4 Function Block Diagrams (FBD).

There are two different kinds of program units:

Functions and Block Functions.

The functions are the so called "procedures" that can have many inputs and only have one single output.

A Function Block can have several inputs and produce several outputs when executed.

The manufacturers have developed a software a user can use to access a library, create his own functions or function blocks and using other tools. Since this is a graphical methodology for programming, the blocks are linked together by pointing the output and/or input the user wants to join.

The blocks can be put in any position on the worksheet.

This method is becoming very popular among new and advanced programmers. Most of the manufacturers have included some of the features of this technique in their own control programming software.

10.5 Sequential Flow Charts (SFC).

Since a good number of control processes or production machines operate on a sequential mode and just a few processes working on parallel, the Sequential flow charts is a very powerful programming method. It offers the following advantages:

a) Easy to understand for other disciplines different from electric and electronics.
b) Easy to debug since the program shows the current execution point.
c) More efficient use of CPU or processor resources.
d) The same timer or counter can be used more than once if they are not used in different places at the same time.
e) Failure or wrong operation diagnostics are quite easy to determine.

It follows certain basic rules:

1. Try to split up the operation of the machine in several steps.
2. There are new intructions like "Steps" , "Transitions", SET and RST and the remaining elements you already know.
3. The STEPs can be "Initial" (Checked when the PLC is energized) or "Normal" (the rest of the steps).The PLC only checks (SCAN) the active steps, which are the steps that are operating.

4. To jump from one STEP to another , you use transitions. These ones are logical conditions to be met before the jump. For instance, a timing, a contact, a counting, a limit switch , etc. There is always a transition from an Origin Step to a Destiny Step. After the transition , the origin step is deactivated and the Destiny step is activated.
5. On every STEP, you can activate outputs (with or without a condition). After a transition you don't need to deactivate the outputs since, at the moment the STEP is deactivated, the outputs go OFF.
6. The instructions SET and RST are used for two main purposes: If SET is used with outputs, these ones will remain active no matter if the step is deactivated after a transition and no matters if the condition is not true anymore.

   If SET or RST are used with STEPs , the STEP is activated or deactivated from a different any part of the program without needing to jump to it. Some authors can call this as GSTART (SET) o Gkill(RST) of procedures or steps.

10.6 Flow Diagrams (FD).

There is another not so popular programming method named Flow Diagram, where you can make the program in a graphic way using the known programming tools of System engineers.

Not so many technicians and engineers are familiar with this kind of programming and seem reluctant to incorporate Functions in their programs. On the other hand, it has opened a door to new disciplines, other than electrical, to develop control programs. The new industrial automation is more universal to all kind of professionals.

In general, a complete control program can be written using any number of methods. Every manufacturer tries to develop tools to make the programming labor easier and tries to include as many libraries as possible so the programmer can make more difficult tasks.

Questions

1. What is the IEC61131-3 standard?
2. What is another name for the Instruction List programming?
3. How can a Structured Text ladder programming be written?
4. What kind of working units can be used in Function Block Diagram programming?
5. What are the advantages of programming through Sequential Flow Diagrams?

Notes

WINTRILOGI PROGRAMMING® AND SIMULATION SOFTWARE

11

This chapter will teach you how to use the programming and simulation software. If you have the PLC trainer you will learn how to download the programs to the PLC for real functional testing.

Either you want to learn Ladder Programming through the Simulation software or the complete program that allows you to connect to the PLC trainer; it's a very good idea that you understand all the features and possibilities.

The software Wintrilogi® was developed by Triangle Research Limited as a part of their product line.

This software is being used by many educational institutions around the world. Even more, some engineers test their programs with this software package before writing it for another brand.

In order to write a PLC program that really works, you need to perform the following steps:

1. Have a clear description of the program you want to develop. Try to use your own words to describe how the operation of the control program is going to be. When you write your ideas, the process will lead to a clear definition and understanding of the control tasks that are really required for a safe and continuous machine or process operation.
2. Write the program, using all the elements and programming tools.
3. Compile the program.
4. Solve all the programming mistakes and go back to step 3 and compile again.
5. Simulate your program.
6. Edit and add the required rungs until the program really works as specified.
7. Download the program to the PLC trainer.
8. Make the proper connections of external devices to the trainer.
9. Test the program operation, analyzing the overall performance.
10. Assemble all the real components in a control panel and wire all the inputs/outputs to the real devices.

11.1 Installing the Software.

Important:

In order to get a DEMO license to practice the ladder logic lessons , please send an email to **plctraining@hotmail.com** and **sales@latin-tech.net** including your info and the serial/ receipt number of the book you have bought. For a real PLC trainer check page 403.

You can also visit www.lt-automation.com in case that you want to buy a PLC training station to have real hands-on practice.

To install the software you need to be very careful with three important steps:

First remove any Java version you have on your computer, then install the JRE (Java Run-time Environment) and finally the Wintrilogi® software.

This is one of the main problems when trying to run the software since improper installation will turn in faulty software operation. Try to use the latest JRE software or the one that comes along with the software package.

Please check if you already have a JRE version installed. If you do, uninstall it, using the tool to "Add or Remove Programs" from the Control Panel on your PC or Laptop.

11.2 Running the Program.

Once you have installed the software successfully, select the Wintrilogi application, following the path shown below:

Select the program

Make sure you select E10 Series. If you select H-series you program won't work on the PLC e10 or the PLC trainer PTSE10. However it will work for any program you write: The H series is a PLC family with more input/outputs.

PLC size selection.

11.3 Communicating with the PLC trainer.

If you don't have any of our PLC trainers (check the appendix), please skip this part.

On the menu "Controller" select Serial Port Setup to check that your communication parameters on the PC are matching the ones you have on the PLC.

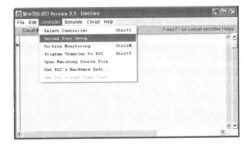

Program the serial port on your PC

On Command String type:

IR*

If everything is OK, you should receive a response string of IR01* or IRXX* where XX is the ID number of the PLC you have, in case that you are programming one of the PLCs on a Network.

Checking connection between PC and PLC trainer.

If you get the following message: "(Warning: No Response from the PLC!)" then you need to check for other problems:

a) Communication port. Verify if you are selecting the right port. This is very important when you are using a USB to serial converter. Not all of them work right away. They usually assign a different serial comm port number.

b) Communication settings. Make sure that your communication settings are exactly the same as the ones on the PLC. To make any change you first need to click on "Close Port", make your changes and then click on "Open Port".

Power on. If the wall adapter is connected you should see that the green LED on the PLC is ON. If not, check the Main Fuse and the wall adapter connection. Once you are connected you can close the "Serial communication Setup and Test" screen.

11.4 Available Menus.

The Menus are very similar to the ones you find on any Windows application .If you are familiar with any other software package you won't have any trouble in using Wintrilogi®.

Once you open the application you will see the following Menus:

File, Edit, Controller, Simulate, Circuit and Help.

11.4.1 File Menu.

This option allows you to handle all the program files.

New (Ctrl+N).

This option allows you create a new program. If you don't use any name it will use "untitled" as default.

Save (Ctrl+S).

Save any changes to the original file. Be careful, to keep the original program intact; better use the option Save As.

Open (Local Drive).

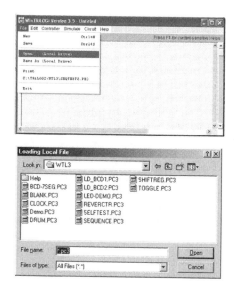

Directory where the Program is going to be saved.

To open any PLC program available on your computer. The simulator can use programs with extension .PE3 or PC3. The ones related to the PLC trainer are PE3.

Save As (Local drive).

To save the modified ladder program files with a different name. This is very useful to generate different versions based on an original ladder program.

Although the name of the program can have a lot of characters, when downloading the file to the PLC Trainer, the name will be cut short to only six characters.

Print.

This command allows you to print the information of your program according to your needs:

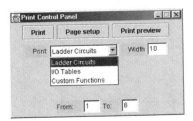

Software printing options.

Page setup button will allow you to make adjustments to the page in size and width.

Print preview button will show how the document would look before printing it.

Exit.

To close the program. You will be prompted to decide whether to save the program or not.

11.4.2 Edit Menu.

This option allows you to make original programs or to make changes to the existing ones.

Abort Edit Circuit.

This leaves the rung with no change if some changes have been made and you have not pressed ENTER yet.

Undo (Ctrl+Z).

Every time that you use this command the ladder circuit goes to its previous modification (maximum 10 steps).

Cut Circuit(Ctrl+X).

To remove circuits. All the circuits , selected by the option box up to the right are Cut and put in the Clipboard. You can paste them somewhere else in the program.

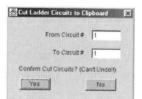

Be careful with this command, once you cut a circuit you can't undo changes.

Removing rungs on a ladder logic program.

Copy Circuit (Ctrl+C).

This command makes a copy of the rungs you have selected in the dialog box, leaving the original circuits intact. The copied rungs are put in the Clipboard.

Paste (Ctrl+V).

After you have "Cut" or "Copied", you can paste all the circuits previous to the current circuit that has been selected.

Find (Ctrl+F).

To locate any particular element with a label name entered in the following dialog box.

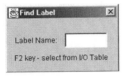

Find a particular element.

Goto (Ctrl+G).

This command allows you to move the cursor, within a ladder program with many circuits, to the circuit you want.

I/O Table (F2).

This table contains all the elements you can use when writing a ladder program.

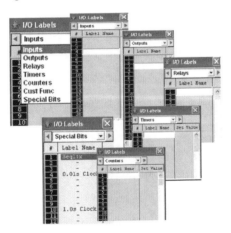

Different tables used to define the elements you plan to use in your program.

Notice that for Inputs, Outputs and Relays you need to enter just the name of the element. For Counters and Timers in addition to the name, they require you to enter the setup value in order to work properly.

The Special bit table allows the user to have access to the internal clocks and the Sequencer.

Customer Functions (Cust Func) is only applicable to our PLC trainers PTS T100 and PTS F100.

I/O Element Identification Letter (F3).

This is a small letter which appears at one side of any contact and helps to identify what element it belongs to:

> i=input, o=output, r= relay,
> t= timer and c=counter.

By typing F3 you can toggle the action of appearing and disappearing the letters.

11.4.3 Controller Menu.

This Menu can only be used if you have the PLC or the PLC trainer. It will allow you to connect the PC to the PLC or PLC trainer.

To connect to the PLC trainer first check the kind of communication port your Desktop or Laptop has. If it's an RS232 port just plug the communication cable that comes with the trainer, otherwise if the port is a USB type, you need to get a **USB to Serial converter.** Be careful when ordering this device, some of them could not work properly.

Select Controller (Ctrl-I).

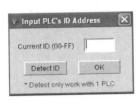

Selecting your PLC's ID

Normally you are connecting your computer to only one PLC or one PLC trainer. Unless you have changed the PLC's ID before, it will be 01. The Detect ID button will allow you to know the ID from the PLC, but it will only work if you previously have made communication with the PLC and if you are not connected to a network of PLCs or PLC Trainers (This is possible, since all of our PLC trainers allow networking).

Serial Port Setup.

You can setup all the communication parameters of your PC's serial port in order to match the ones on the PLC trainer. To do this, you must first close the port, using the button for that purpose. Once you have made all the changes you must open the Port again.

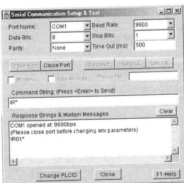

Setting the parameters of the PC´s serial port.

The Command string is a field you use to test if you have communication between the PC to the PLC or PLC trainer. If you send IR*, the PLC trainer will respond with IR01* (Or the current ID number it has).

You only need to do this checking until you are connected.

In case that you have more than one PLC or PLC trainer on a network, you must not use the IR* command because all the devices will respond at the same time, leading to a communication problem.

In order to change the ID of the PLC or the PLC trainer you can use the button "Change PLCID" .

On-Line Monitoring (Ctrl+M).

This menu option allows the user to see what is happening with the status of all the elements, providing two control paths: either the real devices connected to the PLC or forcing element status from the PC screen .

To have On-line monitoring of your PLC trainer you must first establish communication.

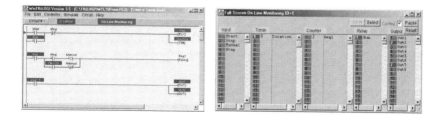

Screens you can see only when you are connected to the PLC.

There are five main columns that contain all the elements used on the project. The left numbers in each column correspond to the specific connection number on the PLC. For instance Start is physically connected to the input #1 of the PLC trainer. When the number is enhanced by a red color or background, it means that the element is "active" or "on".

To allow the status change of any element you must first check the box named "Control" on the upper right corner.

By clicking on any element you can change the status of the element. For example , if you click with the left button of your mouse on Start, this input will be active only for the amount of time the mouse is pressed, once you release your finger from the mouse the input will go back to its original condition. On the other hand, if you use the right button of your mouse, the input will go to the opposite status: If it was OFF it will go ON and will remain like that. If you click once more, then it will go off again. This is what is called a Toggle function.

There are four more buttons which are explained below:

View: Only active for the trainers PTST100 and PTS F100.

Select: If you have a PLC network it will establish online monitoring with a particular PLC , defined by its ID.

Reset: Resets all the data for I/Os and leaves the PLC in original condition.

Pause: Halts the PLC. The ladder program execution is suspended.

The online monitoring can also be performed on the ladder program screen. The activated elements are shown in blue. A highlighted N.C. contact means that the contact is open and a highlighted N.O. contact means that the contact is closed. Be very careful when interpreting the contact status since it leads to confusions.

Program Transfer to PLC   (Ctrl+T).

When your program is ready and has no bugs, it means that compilation has no problem, you can download the program to the PLC or PLC trainer.

You will be asked to confirm if you really want to download the program to the PLC, since it could lead to risks.

Open Matching Source File.

Using this command will inform you about the program that is in the PLC trainer and will open the corresponding program file that has been saved on your PC. It only checks for the name and not the content.

Get PLC's Hardware Info.

When the PLC trainer is connected to the PC, using this command will provide all the available PLC resources: model number, the maximum of input, outputs, relays, timers and counters.

This is the same info you get when you try to transfer a program to the PLC.

10.4.4 Simulate.

This menu is for testing, debugging and simulating the ladder program.

The simulation is an excellent tool to practice on ladder.

WinTRiLOGI® automatically compiles a ladder program before activating the simulator. If an error is found during compilation, the error will be highlighted and the type of error is clearly reported so that you can make a quick correction.

Run (All I/O Reset) (Ctrl+F9).

This option resets every input and output (off state) and immediately opens the simulator. The reason for this is to make sure that you start from the power-up condition on the PLC. The outputs are assumed to be connected to real devices.

Run (reset Except i/p) (Ctrl+F8).

This option preserves the status for the inputs that you had previously set ON or activated. For example if a sensor must be detecting a device before starting the machine , then at the moment you re-start, the input signal must be on. The rest of the data is cleared. After the command is executed then the simulation window is opened.

Continue Run (no reset) (F9).

The purpose of this command is to continue the simulation

from the previous time it was closed. The advantage is that it keeps all the status and values of the elements. If you have made even the slightest change to the ladder program, then the whole program will be recompiled to look for bugs before running.

The first scan pulse (1st. Scan) is not activated again, because it's assumed that the program is continuing from a previous simulation. The main reason to use this command is that you save time, when looking for a simple bug, instead of having to run the whole program and setting the inputs, you just continue from the point of testing you currently are.

Compile Only (F8).

When this command is executed, a new analysis of the resources is obtained. You can see how big your program is.

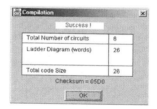

Screen for compilation or problem detection of your program.

Reset All I/Os (Ctrl-R).

All the inputs and outputs are reset without opening the simulator. This helps on the ladder program analysis because of the highlight used to show activation. When the command is executed all the highlights on elements are removed so you can see the original status.

11.4.5 Circuit Menu.

This menu includes some options related to the Rungs.

Insert Comments.

You can add comments that help to understand the operation of every rung or, in some occasions , to help others to know your ideas when writing the program. The Comment editor allows you to copy, cut or paste any text.

These comments are not compiled nor downloaded to the PLC memory and are only stored in your computer file.

The comments are optional and it's up to you if you want to add them or not.

To edit a comment rung just press the Space bar of your keyboard.

Since they are considered like a rung , you can use the circuit option to delete them.

Insert Circuits.

This command helps you to insert a new circuit (or more) just before the currently selected circuit. It's very useful when you want to make changes to the program and want to keep it close to other related elements.

After inserting a rung then the remaining ones are re-numbered.

Move Circuit.

This allows to relocate rungs in the ladder program. First select the circuit you want to move, then select the command "Move Circuit" and finally select the destination location.

If you plan to move several circuits ,use Cut Circuit and Paste Circuit from the Edit menu instead.

Append Circuit.

To add a new circuit to the ladder logic program right after the last program circuit.

Delete Circuit.

Use this command carefully since it's no possible to UNDO the result. If you want to delete one or a range of rungs just fill the data on the window.

11.4.6 Help Menu.

This menu option contains most of the information of this manual. However, since this is a book , we have added some extra comments to make it clearer and useful.

11.5 Loading a program you have written before.

Follow the path as shown on the image below:

FILE>OPEN (Local Drive)>WTL3>Your File

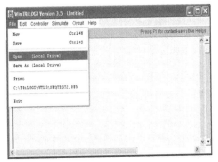

Screen to open an existing ladder logic program.

For this case, we want to open WATERLVL.PE3 so, once it's highlighted with the mouse, click on Open.

The PLC program corresponding to WATERLVL.PE3 will open.

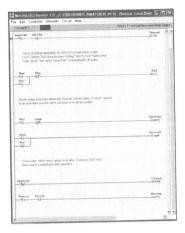

Program opened and ready to be used.

11.6 Simulating a program.

If you want to simulate the project prior to downloading it to the PLC, click on "Simulate" and then "Run". Remember that to perform a Simulation of a ladder program you do not need to be connected to the PLC or PLC trainer. In fact, you don't need the hardware at all.

Screen to select software program simulation.

A new screen will appear, showing the status of most of the elements. If you want to simulate the activation of any input, just click on that particular input. Remember tht clicking on the left and right buttons lead to different results.

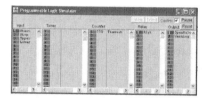

Screen to force or visualize elements´ status.

If you don't want to perform any simulation, just skip this step.

11.7 Transferring a program to the PLC or PLC trainer.

If you don't have the PLC or PLC trainer, just skip this part.

To transfer the program to the PLC or PLC trainer Select "Controller " and then "Program Transfer to PLC". You can also use the keys "CONTROL" and "T".

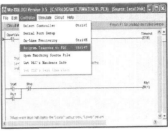

Screen to transfer the control program to a real PLC.

A small window will show up, prompting you to enter the ID (if you know it) otherwise click on "Detect ID" and it will get the ID for you.

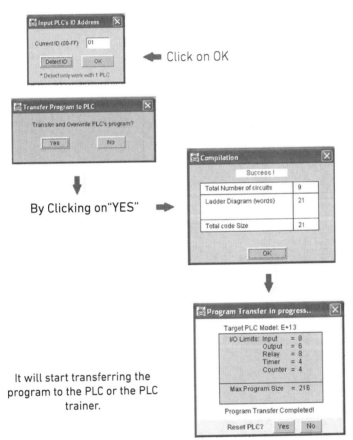

Click on OK

By Clicking on"YES"

It will start transferring the
program to the PLC or the PLC
trainer.

11.8 Online monitoring or control.

If you want to see the ON-LINE operation of the PLC program then select: Controller>On-Line Monitoring. You can also use the keys "Control" and "M" to go to On-line monitoring mode, as shown on the screen below.

Selecting Online Monitoring on the screen.

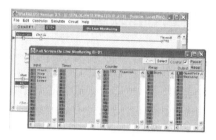

Pop-up screen with all used ladder elements.

Then the screen of On-line monitoring will show up. Now you can command the PLC either from the real push-buttons and switches on the trainer or the inputs on the screen of your PC. Both options will have the same effect on the PLC program.

11.9 Using the push buttons, switches and relays on the PLC trainer.

The different devices on the trainer are connected to the PLC in the following way:

INPUTS.

Device	Is connected to the PLC input #
Pushbutton 1	1 or IN1
Pushbutton 2	2 or IN2
Switch 1	3 or IN3
Switch 2	4 or IN4
Switch 3	5 or IN5
Switch 4	6 or IN6

PLC trainer's real input devices connection.

As an example you could say:

On the PLC trainer the Pushbutton 1 is connected to the Input 1 of the PLC. It would be the same either clicking on input1 through the software simulator or pressing down the Pushbutton1 on the trainer to activate the input 1 on the PLC.

OUTPUTS.

PLC Output #	Is connected to	Contact terminals
1	Relay 1	Out R1A and OutR1B
2	Relay 2	Out R2A and OutR2B
3	Relay 3	Out R3A and OutR3B
4	Relay 4	Out R4A and OutR4B

PLC trainer's output terminals.

As an example:

On the PLC trainer the outputs on the PLC are connected to external relays. Contacts with 7 Amps current capacity are available on the terminal strip.

The activation of Output 1 on the PLC, will energize the Relay 1, which in consequence will close the contact with terminals OUTR1A and OUTR1B.

There is a fuse protecting every relay's contact that will blow out when more than 7 amps are passing through them.

11.10 Writing a program for the very first time.

Let's suppose that we are asked to write the following control program:

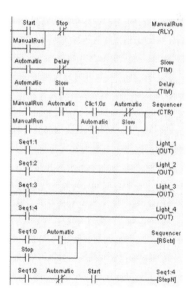

Follow the steps below :

1. Open the program.

2. Select E10

3. Create the name of your program: FirstProg.PE3.

We have chosen this name instead of FirstProgram.PE3 in order to keep the original intact.

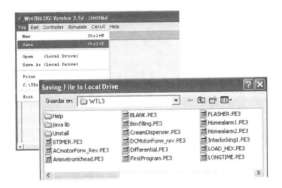

4. Open the I/O table to enter all the elements you want to use, by pressing F2.

Define the elements that you require: Inputs, outputs, relays, timers, counters and sequencers. You don't need to worry if you forget or need more elements in the future; you can add them later. The importance of knowing what to use is that you get an idea of the resources you are going to need, either the devices to be connected to the PLC or the number of physical inputs and outputs that can meet the specifications.

Inputs: Start, Stop and Automatic.

Outputs: Light1, Light2, Light3 and Light4.

Relays: ManualRun.

Timers: Slow , Delay.

Counters: Sequencer.

Sequencer: Number 1, the only one that this PLC model has.

The Inputs, Outputs and Relays are a straight forward procedure by filling out the corresponding I/O table.To enter the timers you need to enter first the name and then the setup value. From the graphic you see that these values are 17 and 10 respectively. It means that the real time value is 1,7 seg and 1,0 seg.

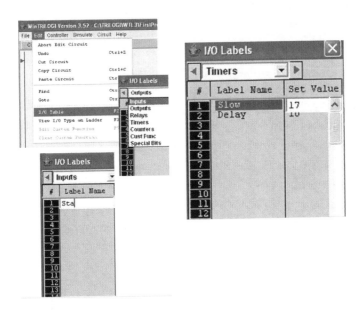

The counters also require a set value. First enter the name then the maximum value it will count. In this case the counter is associated to the only sequencer the trainer has, this is why we call it Sequencer, but I could have been a different name.

The sequence will have 4 steps or the same, the counter will only count up to number 4.

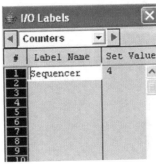

5. Enter in Edit mode by pressing Spacebar. You will see that a new graphic menu will appear on top. You can also "double click" on the busbar in front of the small read triangle.

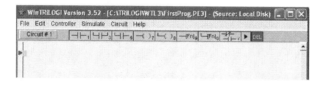

You will know that you are in "Edit" mode because there is a yellow square in front of the red triangle and that there is a new menu bar. The circuit or rung that you are about to add or edit's also shown.

6. Click on the Icon 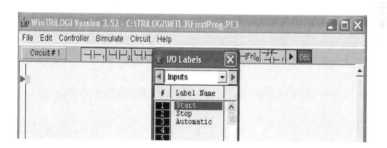 to add a normally open contact. The I/O table will appear. You can notice that there is a small number, in this case 1, that you can select to obtain the same results of clicking on the icon.

If you get the Input table, click on Start and then you should get the following screen:

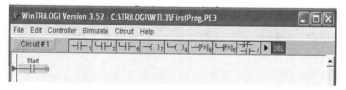

The I/O table hides automatically. You can see that a "Normally open" contact from "Start" has been added.

7. Click on the Icon  or press1 again to add a normally open contact from Stop, following the same procedure used in the previous step.

8. In order to change this last NO contact to NC, click on 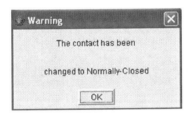 (or selecting the key of "/") , then click on OK . The resultant Screens will be like:

9. Now we need to add an output. This can be done by clicking on 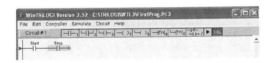 (or pressing the number 7)and selecting the ManualRun from the Relays in the IO table.

Once the ManualRun relay is selected you obtain the following screen:

10. Now we need to add a NO contact from ManualRun in parallel with Start. Before going any further you must click on Start.

Once it´s highlighted you can click on 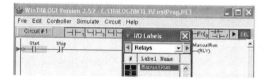 or simply press the number 3.

Since the contact we must add is a contact from the Relay named ManualRun just select it, to obtain the following screen:

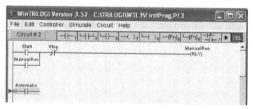

11. Then we are ready to write Circuit #2. Just click a little below the contact we just placed or simply hit Enter.

Notice that the yellow square is indicating circuit #2. Click on the Open Contact Icon and then, from the Inputs select Automatic.

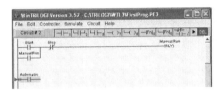

12. To add a NC contact from the timer named Delay in series with the NO of Automatic, Click on the Icon 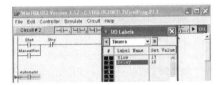, then scroll down in the IO table until you find Timers and select Delay.

A new contact from the timer has been added:

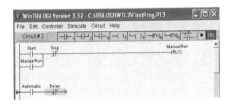

Click on 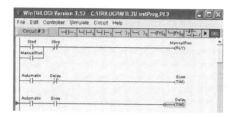 to change it to NC, since by default it´s a NO contact.

And finally, to add the timer we click on the icon ⊣ ⟩₇ , scroll for timers and select the timer Slow.

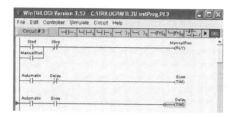

With the same procedures, add the next rung in order to obtain circuit #3 completed.

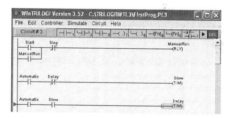

In this case the NO contact from Slow remains without change.

13. This part is quite important since we need to know how to add combinations of contacts.

First add two NO contacts of the relay Manual Run and the Input Automatic as shown below:

Click on the contact ManualRun to highlight it again and then click on the icon  to place the parallel contact.

Notice that to the left of Manual run contact there is a blue cross, and that the icon has changed to ⊣⊢ with a yellow background to prompt you to define the other element you want to make the connection.

So, just click on the Automatic contact to highlight it and then on ⊣⊢. From the I/O table scroll for Relays and then select ManualRun.

You shall have the following screen:

Click on the contact named Automatic again to highlight it and then use the right arrow of your computer to move the cursor to the right. A small yellow square will appear on the right of the contact as shown next:

Then add a new contact by clicking on , and then scroll until you find Special Bits on the I/O table. Select 1.0s Clock.

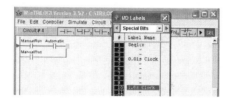

The Contact from the clock will appear. With the same procedure add the NC contact from Automatic until you obtain:

With the left arrow move the cursor to highlight the clock contact and then click on the icon ⊣⊢. Using the same procedure as before, highlight the contact named Automatic , click on ⊣⊢ and then select the input Automatic , from the I/O table.

To add the contact from the timer named "Slow", just click on and scroll for Timers , then select "Slow".

In order to be able to add the output, click on the NC contact named Automatic and move the cursor to the right. A small yellow square will appear on the right side of the contact as shown.

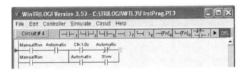

According to the original program we need a counter with the name Sequencer. Since the counters can be obtained as an output from the I/O table you only need to click on ,look for Counters and select Sequencer.

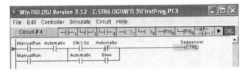

14. Circuit #5 requires a special instruction or contact from the sequencer. To start just hit Enter or click below the current circuit.

You should see the red triangle and the yellow square to make sure that you are in edit mode, otherwise click twice on the rung or press Spacebar.

Click on the icon 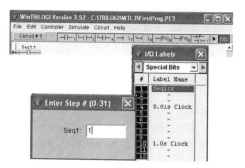 and look for Special Bits then select Seq1:X. You will be prompted to enter the Step number, which is "1"in this case.

With 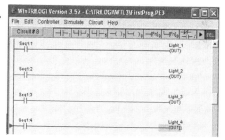 add the output Light_1. Continue adding the new circuits referring to the sequencer's steps 2, 3 and 4 and its corresponding lights :

Light_2, Light_3, Light_4.

15. To create Circuit #9 the only part we don't know is the instruction on the right, if you have read all the above steps you can write the program up to here:

Notice that there is a yellow square at the right side of the NC contact denoted as Automatic. You only need to click on  or simply press the number 9 on your PC's keyboard.

Select the command Reset Counter and then select the Counter named Sequencer.

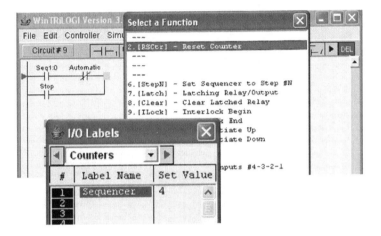

The resultant Circuit#9 is as follows:

16. The Circuit#10 can be obtained with the same procedure. To obtain the last instruction StepN, you also need to click on .

Enter "4" , when prompted.

The final result and last circuit of your program will look like the one below.

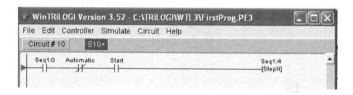

It´s recommended that you periodically save your program. If you have not done it yet, do it through the File menu.

Now you can proceed with simulation or downloading. These procedures are in other previous readings.

Questions

1. What do you need to do first when installing the simulation and programming software Wintrilogi?
2. How do you test your communications with the PLC or PLC trainer?
3. How do you go to Edit mode on the software?
4. What do you need to do to simulate your program?
5. How do you transfer the program in your PC or Laptop to the PLC or PLC trainer?
6. What do you need to do to go to Online Monitoring?
7. How do you delete a rung (circuit)?
8. How do you insert a rung (circuit)?
9. When using the simulation software, what effects can you have on a input if using the left or right mouse buttons?
10. How do you put the Set value for a timer?
11. How do you get the StepN instruction?

12. What do you do to insert a comment on a particular rung (circuit)?

Notes

LESSONS ON PROGRAMMING

12

This chapter contains 40 lessons to guide you through the mastering of the instructions. All the lessons are practical so you need to write the ladder program, simulate it by software and then download it to the PLC Trainer (if you have one).

How to practice the PLC lessons.

In order to have a good understanding of PLC programming, we strongly recommend the following steps:

1. First read all the theory of previous chapters.
2. Read all the lesson and try to understand the program.
3. Type the program always, no matter if the program looks like anyone from other previous lessons. Practicing on writing programs is very important.
4. Compile the program and make corrections for any mistake you had made.
5. Simulate the program through the software and based on the results, try to answer the questions.
6. If you have the PLC trainer, transfer the program to the PLC and for simulation use the real push buttons and switches available.

7. Save every program you make with a meaningful name
8. Add your own comments to the rungs according to your discoveries.
9. Write about your experiences with every lesson.

Lessons.

Lesson 1.   Basic Circuit with NO input contact.
Lesson 2.   Basic Circuit with NC input contact.
Lesson 3   One input driving multiple outputs.
Lesson 4.   The same input with two different contacts.
Lesson 5.   Two NO inputs connected in series.
Lesson 6.   Rungs with several inputs.
Lesson 7.   Rungs with NO and NC contacts.
Lesson 8.   Several activation paths on the same ladder rung.
Lesson 9.   Basic circuit with lock.
Lesson 10.   Basic Circuit with Stop.
Lesson 11.   Adding an Internal Relay.
Lesson 12.   Timers.
Lesson 13.   Timer with a NC contact.
Lesson 14.   Timers in Cascade.
Lesson 15.   Timers .Pulse generation.
Lesson 16.   Timers. Train of pulses.
Lesson 17.   Counters. Unlimited counting.
Lesson 18.   Counters. Cyclic counter.
Lesson 19.   Counters. Driving an output.
Lesson 20.   Counters. Independent resetting of a counter.
Lesson 21.   Counters. Using internal bits or clocks.
Lesson 22.   Counters in cascade.
Lesson 23.   Counters. Driving an output.
Lesson 24.   Timers and Counters.
Lesson 25.   Timers and Counters.
Lesson 26.   Sequencers.

Lesson 1. Basic Circuit with NO input contact.

This is your first experience with ladder so, let's try to start with a basic circuit as shown below:

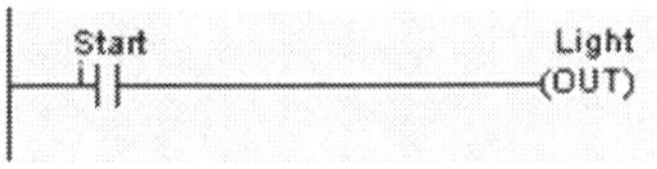

Start is an external push button connected to the first input and the Light is one light bulb connected to the first output of the PLC. For now, we don't need to worry about connections.

When you press the button, the contact Start in the PLC program will close and, as a result, the output Light will energize as well.

The light will be On as long as you have your finger pushing the button down.

Questions.

1. How can you use a Push-button, a toggle switch and a level switch as an input?
2. How fast the output becomes active?
3. Could you have more inputs or more outputs in parallel?
4. Could you have more inputs or more outputs in series?
5. How many instructions is the PLC executing here?
6. Could you suggest an application for this circuit?

Notes

Lesson 2. Basic Circuit with NC input contact.

This lesson explains how to use a NC contact with a NO pushbutton connected to an input.

For this circuit we have changed the NO contact, from the input named Start, to a Normally Closed one. Because of this normally closed condition, there is nothing preventing the activation of the Light. Assuming that we are using a physical NO pushbutton (because we could use a NC one) the only possibility of turning the light OFF is when you press the button Start.

The software simulator only allows you to simulate NO pushbuttons, which you close when you click on the screen.

Questions.

1. What happens if we use a real NC pushbutton connected to the inputs?
2. When can you use a circuit like this one?
3. Is there any difference in using real NC and NO pushbuttons or switches?

Notes

Lesson 3. One input driving multiple outputs.

This lesson allows you to understand using of multiple outputs activated by one single input.

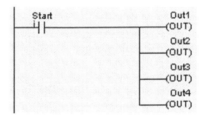

This circuit has one contact named Start and, activating it, its contact will close and energize four outputs. In general you can consider that all four outputs on the right are depending on the activation of Start.

Questions.

1. How many outputs can we connect in Parallel?
2. Are the outputs activated simultaneously?
3. Is it the same to put Out4 on top and then the rest on a non-specific order?

Notes

Lesson 4. The same input with two different contacts.

This lesson helps to understand that you can assign NO or NC contacts to an input and how they work.

```
         In1                          Out1
        ─┤ ├────────────────────────(OUT)

         In1                          Out2
        ─┤/├────────────────────────(OUT)
```

This short program has two rungs. Each rung has a different output, but they both are controlled from the same input.

The only difference is that the contacts have different normal status on each rung: The first rung includes one NO and the second rung a NC contact.

In general one same input can appear several times on different rungs, with NO or NC contacts. In fact, it's quite possible that you have to use the same contact several times in most of the rungs of your control your program in order to create your control logic.

Questions.

1. Are Out1 and Out2 activated at the same time once In1 is active?
2. For any rung, could we add the same contact in series but with opposite condition (If it´s open then a close contact and viceversa?
3. For any rung, could we add the same contact in parallel but with opposite condition ?

Notes

Lesson 5. Two NO input contacts connected in series.

The main purpose of this lesson is to understand the series connection of contacts. Also known as AND logic gate.

The above circuit shows two NO contacts, connected in such a way that the light will turn on only when BOTH contacts are simultaneously closed. As you remember when this happens, there is "current" flow from left to right.

If you are using the software simulator, since you can't click on two contacts at the same time, change one of the contacts to a NC condition.

If you are using a PLC trainer, you can use the first two inputs (that are connected to pushbuttons) and press them simultaneously.

Questions.

1. Can one input be activated first and then the other?
2. Can you use one pushbutton for one input and a switch for the other? You can use the trainer for this (In1 and In2 are Pushbuttons , the remaining inputs are connected to switches.
3. Can we put the Emergency contact first and then the Start contact ?
4. Can you really activate the inputs on simultaneous way?

Notes

Lesson 6. Rungs with several inputs.

This lesson helps you to practice about using multiple inputs and how they interact to create several possible paths to activate an output.

This ladder circuit shows two possible paths to energize the output Out1, either you use In1 and In2 or you use In3 and In4.

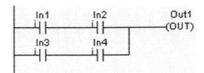

Notice that you have a lot of combinations for the inputs. You can use the mouse´s right button on the software simulator since it only allows you to activate one input at a time. If you have the PLC trainer, you can explore all the possible combinations.

Questions.

1. If we change all the NO contacts to NC, will it be the same?
2. Does it matter if the inputs are connected either to Pushbutton or switches?
3. What happens if all the inputs are activated?

Notes

Lesson 7 Rungs with NO and NC contacts.

This lesson helps to analyze how the activation path works when there is a combination of NO and NC contacts.

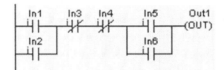

If you are using the PLC trainer, please check that all the switches are not activated (the green lights on the PLC must be off. With the software simulator you don't need to worry about this.

In this rung you can observe that there are four possible paths to activate output Out1:

In1 activated, In3 not activated, In4 not activated, In5 activated
In1 activated, In3 not activated, In4 not activated, In6 activated
In2 activated, In3 not activated, In4 not activated, In5 activated
In2 activated, In3 not activated, In4 not activated, In6 activated

Questions.

1. What happens to the logic if you keep In3 always activated?
2. What happens to the logic if you keep In2 always activated?
3. What happens to the logic if In1 and In6 are always activated?

Notes

Lesson 8. Several activation paths on the same ladder rung.

This lesson has the purpose of studying the number of possible output activation paths when several inputs are used.

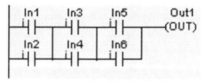

If you are using the PLC trainer, please check that all the switches are not activated (the green lights on the PLC must be off. With the software simulator you don't need to worry about this.

In this rung you can observe that there are four possible paths to activate output Out1:

In1 activated, In3 activated, In5 activated.
In1 activated, In4 activated, In5 activated.
In1 activated, In3 activated, In6 activated.
In1 activated, In4 activated, In6 activated.
In2 activated, In3 activated, In5 activated.
In2 activated, In3 activated, In6 activated.
In2 activated, In4 activated, In5 activated.
In2 activated, In4 activated, In6 activated.

Questions.

1. When all inputs are activated, waht happens to the output?
2. By changing the order of the parallel contact blocks ,do we get the same result?
3. Can any of the inputs be duplicated on the same rung?

Notes

Lesson 9. Basic circuit with lock.

With this lesson you will learn how retention contacts work.

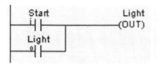

Adding a contact from the same output in parallel with Start, we will have the following sequence of events:

You press the Start button, the Light will activate and since it's active now, then the associated contact Light in parallel with Start will close, providing an alternative path to keep the output activated.

You can remove your finger from the pushbutton and the light will remain on because the contact Light is closed. It's like the contact is replacing the action of you pushing the button.

This will happen so fast that you can assume that all the actions will occur at the same time.

Now we have a problem: The light is ON and there is no way to turn it off unless you power off the PLC.

Questions.

1. What type of contact is the one named "Light".
2. Could you explain how the PLC will internally execute this logic?
3. How fast will the contact "Light" close?

4. Could you provide solutions to turn off the light without powering off the PLC?
5. How many instructions is the PLC executing here?

Notes

Lesson 10. Basic circuit with Stop.

This lesson is helpful to understand the classical Start-Stop configuration. Notice that we have added a Normally Closed contact named Stop . It should be connected to another PLC input.

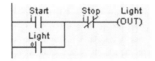

Since the contact from Stop is closed it's like we were having all the elements from lesson 2 and this contact becomes "invisible " for the circuit.

The explanation is as follows:

You push Start, the Light is turned On and it remains On, even if you don't continue pressing Start. This happens because that associated contact from the output Light is closed.

After you have followed these procedures if you press Stop, even for a very short time, the light will deactivate. The reason is that pressing the Stop button, you are breaking the path for the circuit, so it will go to its original condition and will require that you press Start again to re-initiate the process.

Questions.

1. Please explain which of these elements must be connected to physical inputs or outputs.
2. For a given input can we associate a NO and NC contact at the same time?
3. Why is the physical pushbutton contact named STOP?
4. How could we replace the NC contact STOP and use a NO contact instead?
5. If the output Light is driving an external relay, could we use a NO contact from that physical relay to replace the contact with the same name on this circuit?

Notes

Lesson 11. Adding an Internal Relay.

This lesson allows you to understand the similar use of internal and external outputs. For this case we will use an internal relay.

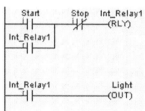

In our last lesson we were using a contact from the external coil to keep the coil energized, once you removed the finger from the push button named Start. In this case we will use an Internal Relay to do the same function . An additional contact from the same internal relay is used to activate the output named Light.

The program will work exactly the same as in the previous lesson, the only difference is the relay type.

Physical wiring requires a lot of auxiliary relays. Using Internal relays also consumes more programming steps than using the direct output.

Questions.

1. What is the advantage or disadvantage of this circuit with respect to the one described in lesson 3?
2. How many internal relays could we use on a given program?
3. How fast is the activation of the output?

4. If you had to make a physical wiring, how could you specify this relay to place a purchase order ?
5. Is it the same if you use two PLC outputs in parallel rather than having a real relay with two contacts, which can close simultaneously?

Notes

Lesson 12. Timers. Basic concept.

This lesson will introduce the concept of Timers in PLCs.

We are using the typical Start/Stop circuit implemented with an internal relay. Notice, that the Internal relay is commanding a timer. The timer's contact is commanding the external output denoted as light.

Once you have activated the Internal Relay (by momentarily pushing Start) then the timer starts counting the time. Right after the timer elapses then the the timer's contact closes, allowing the activation of the external output named Light.

If during the timing process, you press Stop, then everything stops : the timer and the light will turn off.

Questions.

1. What is time base?
2. What is the maximum timing you can use?
3. What is the minimum timing you can use?
4. How many timing contacts can you use?
5. How can we stop a timer?
6. How can we pause the timing?
7. How can we activate the Light with a different option, other than the timer's contact?
8. Could you mention applications for this circuit ?

Notes

Lesson 13. Timer with a NC contact.

This lesson analyzes the application of NC timing contacts and its effect on any driven output.

The example is the same as before. There is only one difference: the timer's contact has been changed from NO to NC.

In this case the Light will be always on and will only go off after we start the timing process, by pressing the Start button.

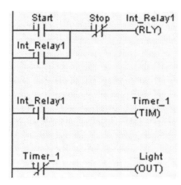

After the timer elapses then it changes the state of the contact. It was originally closed and after the timing, then it will open. Remember that to activate an element on the right ,we need to guarantee the all the contacts on the left are closed.

Questions.

1. How many programming steps is the PLC executing here?
2. Is this timing fixed or variable?
3. Can we simultaneously use another NO timer's contact but with a different output?

4. How can we have the light initially off , using the same operation described here?
5. Could you mention applications for this circuit ?

Notes

___

Lesson 14. Timers in Cascade.

This lesson introduces the concept of using two timers in cascade.

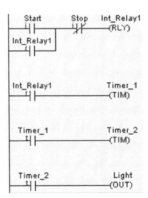

If you take a careful look to the above circuit, you'll realize that Timer_1 is enabling Timer_2. This is what we call a "Cascade".

You can use as many timers as you want and the total timing will be the sum of the individual preset values you have defined for each timer.

Questions.

1. What is the maximum timing that can be achieved with the two timers above?
2. What is the total timer if the preset values of the timers are 10seconds and one minute respectively?
3. Describe the operation for the outout named light if you change the Timer_2's contact from NO to NC.
4. Is it possible to cascade devices other than timers?

Notes

Lesson 15. Timers . Pulse generation.

This lesson will help you understand how to develop a ladder logic program that performs a temporary activation (also called "pulse") for an output.

This circuit allows you to generate a "Pulse". A Pulse is the change of a state for limited time. In this case the lamp is OFF. By pressing Start, the lamp will turn on for some time and then automatically will go off.

There is no need to press Stop to turn off the lamp.

When you have continuous pulses this is called " Pulse Train".

Questions.

1. What is the minimum time we can achieve?
2. What is the purpose for Timer_1 and Timer_2?
3. When is the timing starting? At the moment we press start or at the moment we release the button?
4. What is the longest waiting time we can have? The maximum pulse duration achieved?
5. Where in the program is the pulse being disabled?

<u>Notes</u>

Lesson 16. Timers. Pulse train.

This lesson will help you to practice with the DUTY CYCLE concept. When you need more than one pulse then you can use what we call a Pulse Train.

In this example, the light will be on and off in a continuous cycle until you push Stop. Observe that we have removed the NC Timer_2's contact and we put it on the Timer_1's rung. Since both timers can have different set up, by changing this parameters you can modify the "DUTY CYCLE", which is the relation between the ON and OFF times.

Questions.

1. What is the duty cycle if Timer-1 and Timer-2 are both set to 1 sec .
2. What happens if the timer settings are 1 and 2 respectively?
3. What happens if the timer settings are 2 and 1 respectively?
4. What is the visible effect on the lamp if we change the duty cycle?
5. If we use a different device other than the lamp, let's say a coil, how do we make sure a proper operation?
6. What is the easiest way to have automatic pulse trains?

Notes

Lesson 17. Counters. Unlimited counting.

This lesson introduces the basic aspects of using a counter.

The input Sensor is a physical sensor that detects any object. At the moment the object is "detected" or "seen" by the sensor then its associated contact closes, producing a pulse for the counter.

The counter is previously programmed with a number, which is the one we want to count.

In this example, we are just "counting" and there is no use for the counting information.

Questions.

1. Mention five examples where a counter is required.
2. Does the sensor require any special sensing duration while detecting the object?
3. Why is this process called an "unlimited" counting?
4. How can you re-start the counting?

Notes

Lesson 18. Counters. Cyclic counter.

This lesson will help to understand how to count up to certain number (Preset value).

Notice that we have added a last rung which includes a NO counter 's contact. It enables the instruction or command (RSctr) to perform a counter 's RESET.

Using the simulator you see that the counter starts from the PRESET value and counts down to Zero.

Questions.

1. If the counter is programmed to count up to 10, what happens when it reaches the pulse number 11?
2. Is it the same if we use a NC contact for the counter?
3. Is there any relevant action on the counting?
4. What kind of information could we use from the above ladder program?

Notes

Lesson 19. Counters. Driving an output.

This lesson teaches how to use the information derived from counting process.

Based on the information from the previous lesson, notice that we have added to the last rung, an output named light.

In this case, the counter's RESET and the light will go ON, at the same time, when the counter reaches its PRESET value.

Questions.

1. How many counting pulses are required to turn Light on?
2. What happens if the sensor is not very fast?
3. How can we have the Light on for more time?
4. Could you try to mention an application for this program?

Notes

Lesson 20. Counters. Independent counter resetting.

This lesson shows a different way of resetting a counter, other than the automatic resetting.

As you can observe at the above ladder program , we have an external input named Reset_CNT that will allow us to provide a manual reset to the counter. We don't mind about the current counting because, after pressing Reset_CNT the counter will be forced to reload its preset value.

Remember that the counter will count down to zero.

Questions.

1. Try to give examples when a manual resetting for a counter is needed.
2. What happens if we first push the Reset_CNT button, before pressing Start?
3. What happens if we keep the Reset_CNT button pressed?
4. Loading the counter with 10 as preset value, what happens every time we press Reset_CNT?

Notes

Lesson 21. Counters. Using internal bits or clocks

This lesson will introduce the use of automatic pulse train generators in the PLCs.

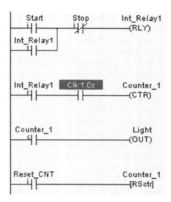

We have removed the sensor input and we added a 1 second clock instead. This means that the counter will receive the pulses from the clock signal, provided the Int_Relay1 is closed.

In other words to enable the counter we only need to press Start.

Questions.

1. Could you provide any example to use an internal above signal?
2. If we change the clock to 0.5 sec how long does the counter takes to reach 10?
3. What's the accuracy of this internal clock / counting?

Notes

Lesson 22. Counters in cascade

This example will show two counters connected in cascade.

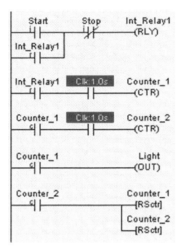

This is a pretty basic cascade connection. Once Counter_1 reaches its PRESET value then its contact (also named Counter_1) will close and will enable the counting for Counter_2. This last one receives the pulses from the clock signal.

Observe that there is a rung which includes a NO contact from Counter_1 driving the Output Light.

A contact from Counter_2 is in charge of providing the reset command for both counters as shown in the last rung.

Questions.

1. When is the lamp turning on?
2. Is this cascade system starting over once the preset value has been reached?

3. If we remove the clock before the Counter _2 and we add a RESET counter output in parallel with Counter_2. What is the new counting?
4. With the original program, what is the maximum counting number that we can achieve?

Notes

Lesson 23. Counters. Driving an output.

This lesson shows a way to combine counters and to use its outputs.

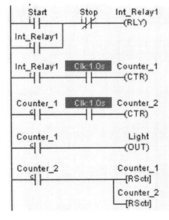

This circuit shows two counters (Counter_1 and Counter_2) counting up to any given preset value. We are using a 1 sec clock to avoid waiting for the signal coming from a sensor.

Once the internal relay is energized, the counting will start counting once per second, when it reaches its preset value then its associated contact will close, providing an enable signal for the second counter. In other words, Counter_2 will work only after Counter_1 has finished.

The on status for the lamp is depending on the activation of the Counter_1's contact.

The contact associated with Counter-2 is used to provide the reset signals for both counters and to start the process all over again.

Questions.

1. If the preset values are 10 and 5 for Counter_1 and Counter_2 respectively, how long does it take for the lamp to turn on?
2. For the same preset values of the counters. How long does it take for the lamp to turn off?
3. Is it the same counting 10 pulses of one second than loading a timer with a PRESET value of 10 seconds?

Notes

Lesson 24. Timers and Counters in combination.

This lesson explains the way we can use a combination of timer and counter to get the same results.

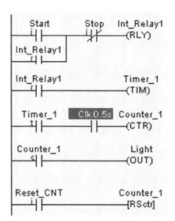

We can combine timers and counters to provide the same result as the one obtained by having just two counters.

When the counter's input is a periodic signal (in this case an internal clock) we can easily replace the counter by a timer, making the proper calculations (Counter's preset value multiplied by the Clock signal).

In this example, the Counter_1 is enabled once the time of Timer_1 has expired.

Questions.

1. What is the total time, from the moment the timer starts upto the moment the lamp goes off ,if the timer is programmed with 5 seconds.?
2. The same but when the counter is programmed with 10.

3. Does the lamp turn off automatically?
4. What do you need to do to put the lamp to blink once per second?

Notes

Lesson 25. Timers and Counters in combination again.

This example shows how to use both, timers and counters, in a combined action to have several timings.

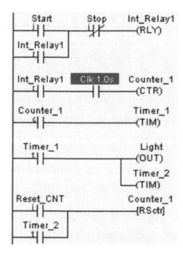

In this circuit we are using one counter and two different timers. Counter_1 is enabling the rest of the timings.

Once Counter_1 has reached its preset value and subsequently Timer_1 has elapsed, the light will go on. A given time later, defined by Timer_2's preset value, the light will go off and the process will re-start all over again.

Questions.

1. If the Counter_1 is programmed with 8. What will be the preset value for a timer to replace the counter?
2. If you push the Reset_CNT button ,what is the effect on the operation?
3. How can we modify this circuit to make the transition (lamp on – lamp off) to occur only once.

Notes

Lesson 26. Sequencers.

The following example explores the basic use of a sequencer on the control of four outputs.

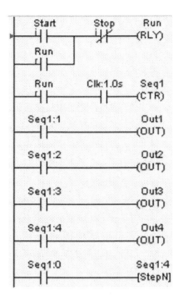

Using a sequencer is something very interesting. Notice that we call the Sequencer instruction using the Counter number 1.It's just because of this particular software design.

Other manufacturers might use different symbols and drawings.

The sequencer is based on a counter named Seq1, which has a preset value loaded with 4.The outputs will turn on only when the respective sequencing step is on.

Questions.

1. Mention the order the lamps will turn on and off.
2. Let's say that we want the output 3 on, when the sequencer is on step 3 and 1. Is it possible?
3. What element does determine the sequence speed.
4. When would you use a sequencer instead of ladder logic?

Notes

Lesson 27. Sequencers. Visual effects with several outputs.

The use of sequencers is something very applicable in the industrial world.

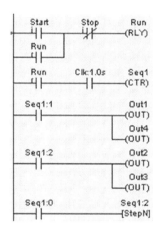

With the sequencer we can activate several outputs at the same time on every sequencing step. For the above program Out1 and Out4 will energize when step 1 is active, Out2 and Out3 will do the same when step 2 is active.

Notice that we only have three steps here, so once the counter Seq1 reaches zero we have to re-start the sequencer by positioning it on Step2.

Questions.

1. In this program the "lights" seem to move from the extremes toward the center. How can we have them moving from the center to the extremes?
2. Here you don't have Out4 and Out5. How can we insert the needed rung?
3. How can we add a different effect?
4. How can we manually stop the sequence on a specific step?

Notes

Lesson 28. Sequencers. Random output activation.

This example shows how to use the contacts of the sequencer steps to create a random effect.

Take a look to the program shown on the next page.

The sequencer can have a lot of steps. The one used by this software is capable of handling up to 32 steps, although we only use 12 in this example.

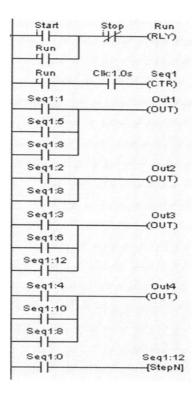

You can make random combinations of step contacts ,as shown above.

The obtained effect is that the outputs will activate in a way that they seem not to follow any specific pattern.

Questions.

1. Try to add more contact steps on the left side of the rung.
2. What happens for the sequencer steps that are not used?
3. Could you suggest any application?

Notes

Lesson 29. Sequencer. Counting the steps and stopping.

This example shows how to control the number of steps to execute when using sequencers.

In this program there is another counter: ItemCount. This counter will be counting the number of executed steps, we want to have for the sequencer, before stopping the process.

Notice that there is a NC contact from this counte,r in series with the stop button's contact.

You can also notice that we have moved the counters' reset to the second rung.

Questions.

1. If we have 8 lights associated to the eight physical outputs, what is the direction of the movement?
2. What are the first outputs that turn on?
3. If the ItemCount is programmed with 10, what are the outputs that will go off before finishing the whole process.

Notes

Lesson 30. Sequencer. Stopping by time.

This example explores another method to stop the sequence: a timer.

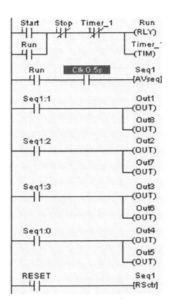

In the example we don't count the number of steps to provide a stop signal. We are using a timer instead of the counter so, once the timer elapses then the process is finished.

Questions.

1. How do we synchronize the preset value of the timer to stop at any given step?
2. What is the purpose of the reset button?
3. For any given value of the timing, what happens if we use a NC contact from Timer_1 in series with Seq 1:2 (removing the other NC contact from the timer which is in series with the Stop contact)?
4. How can we "block" all the outputs, when pushing the reset button, without stopping the sequencer?

Notes

Lesson 31. Using one input to turn on and off an output.

This lesson teaches a trick of using one same input to activate or deactivate an output.

In several occasions we run out of inputs and only one is available so, we need to use the same input to turn on and off the device. Here we are showing two solutions.

The first one uses two important concepts: The "edge" triggered actions that are only executed when the input has a transition from not pushed to pushed (or the opposite). The "latch" and "clear" commands that will turn the output on or off respectively .

The advantage is that the signal only need to be active once (and the logic enabling them) .With the help of two internal relays CR1 and CR2 we "memorize" the last status to provide a "toggle" function.

The other solution is achieved by using a sequencer that is associated to a counter (programmed with a PRESET VALUE of one).

```
Input_2                        Seq1
──┤ ├──                       ─[AVseq]

Seq1:1                         Out2
──┤ ├──                       ─(OUT)
```

Every time we activate the input, a pulse is applied to the counter. The output is always cycling from activation to no activation.

Questions.

1. If we can use the same input to turn on/off an output or a circuit, could we use the same with every stop button?
2. Is it the same for other PLCs?
3. What method is easier to use?

Notes

Lesson 32. Timers and Duty Cycle.

This example considers the application of timers on the calculation and changes of Duty Cycles.

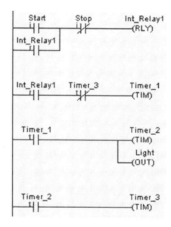

We will use two main timers (Timer_1 and Timer_2) and one auxiliary timer (Timer_3) to understand the Duty cycle. This last one is not really needed but we will use it for illustration purposes.

Timer_1 handles the time we want to have the light OFF. Timer_2 is in charge of the time the light will be ON.

Timer_3 is the minimum time we can wait to start the cycle again.

If the preset value of Timer_3 is really small or minimum, we can neglect its effect on the total time.

Questions.

1. What is the duty cycle when Timer_1, Timer_2 and Timer_3 are programmed with 2 ,2 and 0,1 seconds respectively?
2. What is the duty cycle when Timer_1, Timer_2 and Timer_3 are programmed with 2 ,4 and 0,1 seconds respectively?
3. What is the duty cycle when Timer_1, Timer_2 and Timer_3 are programmed with 2 ,1 and 0,1 seconds respectively?
4. What happens to the duty cycle if we increase the time on Timer_3 to 1 second?

Notes

Lesson 33. Sequencer with timers.

Not all the PLCs include sequencers so we need to use common elements to develop sequences. In this example we will use timers to generate a sequence of outputs.

When we are unable to use sequencers, for any given reason, we can implement a sequential action through timers.

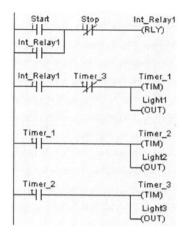

The above ladder program is turning three lights ON in a sequential way. First Light1 turns on, then after a time defined by Timer_1, Light2 goes ON.

The third lamp, denoted as Light3 goes ON after the time for Timer_2 has finished.

Timer_3 has the only purpose of starting the cycle over .

Questions.

1. Do we need to have the same PRESET Value for all the timers?
2. What is the visual effect on the outputs if the preset value for each timer is bigger than the last one?
3. Try to add a fourth sequential lamp named "Light4".

Notes

Lesson 34. Counters enabling counters

This lesson shows how to use the output of a counter to enable a second counter.

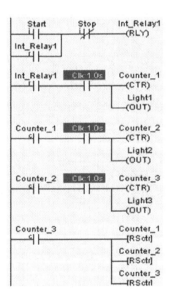

The above program shows a system that turns on a light (in a flashing mode) during a certain time. Counter_1 counts the number of seconds to wait before the lamp starts flashing at a frequency of one time per second. The number of flashes is calculated by the addition of the Preset Values of Counter_2 and Counter_3.

Questions.

For the following Preset values: Counter_1= 4, Counter_2 = 5 and Counter_3 = 1.

1. How many seconds do we wait before the lamp starts flashing?
2. How many times is the Light flashing?
3. How can we change the flashing speed for the output named Light?

Notes

Lesson 35. Sequencer with counters

The purpose of this lesson is to show that sequencers can also be implemented using counters.

We already know that if we use a clock signal along with a counter, it can be assumed that it´s a timer or a "time counter".

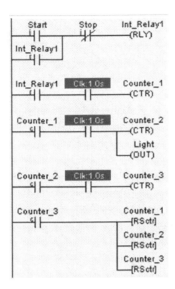

Observe that we are using a one second clock as an input to the counters so, these last ones will be counting seconds according to the PRESET Value they have been loaded with. The three lights will turn on in sequence with a difference of time.

Questions.

1. What is the purpose of Counter_3?
2. If Counter_1, Counter_2 and Counter_3 are loaded with a PRESET Value of 1, 2 and 3 respectively. What is the sequence for the lights?
3. What can you do to have all lights ON for one second, and OFF for another second? The idea is to drive one light at a time.
4. How can you delay the activation of all the lights? Since they are turning ON in sequence, by delaying the first light then the others will be delayed for the same amount of time.

Notes

Lesson 36. Incrementing/Decrementing counting.

The purpose of this lesson is to teach about the up or down counters.

```
   Increment              Counter_1
  ┤ ├                      [DNctr]


   Decrement              Counter_1
  ┤ ├                      [Upctr]


   Counter_1                 Light
  ┤/├                       (OUT)
```

The ladder program shows a counter named Counter_1 which can count Up or Down, depending on which input contact is pressed (Increment or Decrement) and the function applied to it (DNctr or UPctr).

A contact from the Counter_1 is used to control an output named Light.

We will start pushing the input Increment , at the moment the counter is loaded with its SET value and will start decreasing until it reaches zero. At this point the contact associated to the counter will close turning the lamp ON. If we activate the Increment signal, we will have the opposite counting effect.

Questions.

1. Why is the contact Increment driving the function of counter Down (DNctr)?
2. What happens if we start decrementing the counting?
3. What can we do to have the light on for more time?
4. Could we use a ResetCtr instruction? Add this rung to the program.

Notes

Lesson 37. Up/Down Counters

This lesson will allow you to increase or decrease the Present value of any given counter.

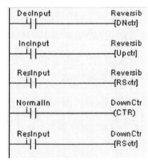

To understand the differences in using normal or reversible counters you can see the former examples. Just create two counters: Reversib and DownCtr. The main difference is that you use the Upctr or DNctr commands for the counter you want to be reversible, and use CTR for the counter you want to be normal. Remember that this last one will count down to zero.

Questions.

1. How many of the PLC inputs are used in this example?
2. Is any of the counters faster than the other?
3. What parameters should you consider to implement faster counters?
4. How many counters can we have or use as Reversible?

Notes

Lesson 38. Using the Latch instruction

This lesson explains the use of two new instructions.

The instructions Latch and Clear are also known as SET and RESET .

They are defined to act on the same output, as the exception to the rules of ladder programming that specifies that an output can only appear once.

The above diagram shows that, using two independent inputs, you can turn on and off the Light

Questions.

1. How long any of the inputs should be pressed to activate or deactivate the output?
2. What is the difference between this one and the common OUT instruction?
3. Could we use just one of the instructions (Latch or Clear)?

Notes

Lesson 39. Using Interlocking

The student will learn how to block parts of the program using the Interlocking concept.

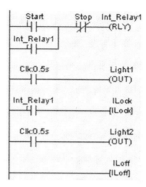

This instruction does not form part of the basic instructions for ladder logic. Some PLC manufacturers have added this instruction as an extra feature to provide better control on the PLC operation.

You can observe that the two special instructions are ILock and ILoff. You can have as many ladder rungs you want between the two lines that include the instructions, and they will only operate or will be executed in the program run, if the contact input, enabling the ILock instruction, is active.

Questions.

1. What happens if we remove the contact Int_Relay1 that drives the ILock instruction?
2. Could you suggest an application of the ILock and ILoff instructions?
3. What happens if we have another ladder rung after the ILoff instruction?
4. Could we add another contact (from any other program element) in series or parallel to Int-Relay1's contact to enable the ILock instruction?

Notes

Lesson 40. Edge triggered events

The purpose of this lesson is to understand the concept of edge triggering.

As mentioned in the theory, the representation of Differential up or down contacts varies from manufacturer to manufacturer. In this case we are using the instruction dDIFD that can be found in the software under Output Coils.

Questions.

1. How does the instruction differential Up operate?
2. How does the instruction differential Down operate?
3. How could we have this edge differentiation in a contact instead of an output?
4. What is the importance of this instruction?

Notes

PROJECTS WITH LADDER LOGIC

13

This chapter will help you to get practice with real application projects. You can write/simulate the programs on your computer or download the program to the PLC trainer for real hands-on experience.

This is a more realistic approach to ladder logic programming and its real application on industrial automation. In order to have a good understanding of PLC programming, we strongly recommend the following steps:

1. Read all the project description.
2. Try to understand the operation and the automation requirements.
3. Take a look at the elements being used.
4. Very important: try to write your own program before taking a look at the one we have.
5. Simulate your program and see if it works properly.
6. Compare your solution with the one we have.
7. Write our program and simulate it.
8. Save every program you make with a meaningful name.
9. Add your own comments to the rungs according to your discoveries.

10. Write about your own experiences with the project in the proper space of the printed material.

List of projects.

1. Box filling system.
2. Sequential Start and Sequential Stop.
3. Home alarm system.
4. Counting of people entering to a place.
5. Automated Ice Cream Dispenser System.
6. Level in liquids with turbulence.
7. Controlling Minimum and Maximum in water levels.
8. Printing machine.
9. Animatronics head.
10. Talking machines.
11. Controlling a DC motor. Forward/Reverse.
12. Starting a tri-phase AC motor with a PLC (Star-Delta connection).
13. Controlling a tri-phase AC motor. Forward/Reverse.
14. Controlling a Stepper motor through a driver.

1. Box filling system.

The pieces are falling down through a slider, first on the left box. After counting 5 pieces, there is a pivoting plate controlled by a pneumatic piston that will route the pieces to the box on the right side. After counting five pieces the plate returns to its original position.

Write a program that can work after a start signal and that can stop at any time, by pressing down a Stop button.

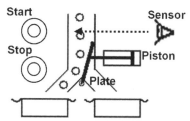

Elements used:

Inputs:

> **Start:** Pushbutton NO, to initiate the operation.
>
> **Stop:** Pushbutton NO, to stop the machine at any time.
>
> **Sensor:** Can be any type of sensor as long as it detects the piece.

Outputs:

> **BoxSelectP:** Drives a two-position valve, which in time, drives a Piston.

Internal Relays:

> **Relay:** to retain the running operation once Start is pressed.
>
> **AuxRelay1:** In conjunction with AuxRelay2 helps to memorize or retain the last status.
>
> **AuxRelay2:** In conjunction with AuxRelay1 helps to memorize or retain the last status.

Timers: None.

Counters:

> **PieceCount:** Piece Counter. This counter is programmed with 5.

Sequencers: None

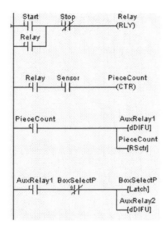

Program Operation.

Supposedly the pieces travel separately thru the slider. This is the main requirement to guarantee that the counting is correct.

To start the machine you press the Start button. This action energizes the auxiliary relay named Relay.

There is a contact from the Sensor, in series with a contact from Relay, to allow that the counting action only works when the machine is running.

When the programmed value of counting is reached (Counter's set value) the PieceCount contact closes to start to actions: Energize AuxRelay1 and reset the counter. Remember that in the case of this software the counting is from Five down to Zero.

We are using the auxiliary relays 1 and 2, activated by differential actions, to remember the last status of the Box Selection Piston (Output BoxSelectP) and change it to the opposite condition: If it was On the next status will be Off and viceversa.

You can notice that the contacts of these outputs are included in the logic of the last two rungs.

Simulation with the Trainer.

To start press the Pushbutton of IN1. To stop at any time or re-start the machine, just press the Pushbutton of IN2.

Slow and continuously toggle the Switch connected to IN3. You will notice that Output 1 (LED1 and Relay1) will activate and de-activate every 5 pulses.

Notes

You can follow the operation directly on the control screen when you are performing ONLINE monitoring.

2. Sequential Start and Sequential Stop.

There are three motors coupled to its respective conveyor belt. After pressing down a Start button, the motors must begin running delayed every 3 seconds, in the following order: Motor 1, Motor 2 and then Motor 3.

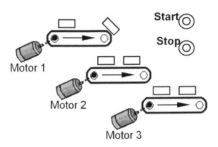

If we decide to stop the conveyor belts at any time, the motors must stop in the same sequence and with the same time separation: Motor 1 stops, Motor 2 stops and then Motor 3 stops.

Elements used:

Inputs:

> **Start:** Pushbutton NO, to initiate the operation.
> **Stop:** Pushbutton NO, to stop the machine at any time.

Outputs:

> **Motor1:** Drives the motor of the first conveyor belt.
> **Motor2:** Drives the motor of the second conveyor belt.
> **Motor3:** Drives the motor of the third conveyor belt.

Internal Relays:

Relay1: Retains the running operation once Start is pressed.
Relay2: Retains the running operation once Stop is pressed.

Timers:

Delay1: Time to start Motor2.
Delay2: Time to start Motor3.
Delay3: Time to stop Motor2.
Delay4: Time to stop Motor3.

Counters:

None.

Sequencers:

None.

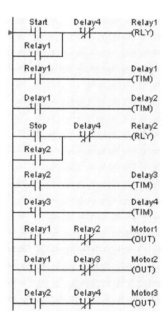

Program Operation.

The idea of having the motors to stop in the same sequence they started before is because you avoid the accumulation of pieces on the conveyor belts.

This program becomes special because this is one of these few cases where the Start and Stop contacts are not located on the same control rung.

You noticed that Start and Stop are the only inputs we have here. We could consider that they both perform the same function in activating its related auxiliary relays and timers. In fact the control rungs for every pushbutton are quite similar.

The ways the contacts interrelate produce the required sequence at the moment of starting or stopping.

For Motor 1 there is no need to use a timer.

Motor 2 obeys a sequence of events. First you press Start, then Relay1 energizes and then it enables the timer named Delay1 which, once the three seconds have run out, Motor2 turns on.

The same procedure can be done to start Motor3.

A very important remark is the fact that the timers remain active (and their related contacts too) after their programmed time has elapsed. This condition allows the operation of the timers that have to do with the stop sequence.

Observe that Delay4 is the one in charge of deactivating both auxiliary relays and , in time, the interlocking contacts of the pushbuttons.

Simulation with the trainer.

You only need to use the RED pushbuttons on the PLC trainer, associated to IN1 and IN2. Since the timers and the auxiliary relays are only accessible through the software, you can use the Online Monitoring option to watch them while they count.

The PLC outputs will get active in sequence: First the Out1, secondly the Out2 and finally the Out3 after you press the first pushbutton. The PLC outputs drive a Relay and each relay closes a contact whose terminals are available on the side connector.

Once the three PLC outputs are ON, means that the three Motors are running, you can press the second button and check that they stop in the same sequence.

Notes

3. Home alarm system.

The alarm system uses 5 sensors distributed among the windows and doors in the house. The alarm will sound if any of the doors or windows is opened for more than two seconds. To prevent false activation, by effects of wind on any window or door, there is a hidden switch that will allow the owner to RESET the alarm once it has activated.

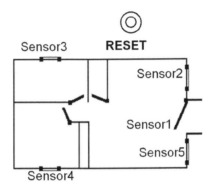

Elements used:

Inputs:

> **RESET:** Pushbutton NO, to re-start the alarm system.
> **Sensor1:** Sensor NO, main door.
> **Sensor2:** Sensor NO, front window.
> **Sensor3:** Sensor NO, side window.
> **Sensor4:** Sensor NO, side window.
> **Sensor5:** Sensor NO, front window.

Outputs:

> Alarm: Siren that sounds when one of the sensors is activated.

Internal Relays:

None.

Timers:

Timer: Time to wait before activating the Alarm.

Counters:

None.

Sequencers:

None.

According to the sensor type (NO or NC) you can write two different ladder programs.

Sensors with Normally Closed contact.

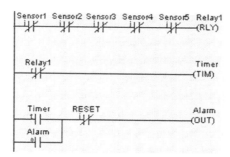

Sensors with Normally Closed contact, that once they are mounted and the doors and windows are closed, will not send any signal to the PLC.

Program Operation.

A switch for a door or a window, has the purpose of detecting when someone is trying to break into the house. These switches can be NO or NC, depending on the kind of action they execute on the signal they send to the PLC. This is why we can write two different programs.

Externally, the inputs of a given PLC can be connected to a sensor that can be assumed as having a NO or NC contact. In the PLC program, no matter the type of sensor you are using you also have the chance of assigning them either a NC or NO contact so, in general, you can consider four different combinations. This can be a little confusing in defining the required ladder logic.

In the first case, when the sensor has a NC contact, the PLC input is active when the door or window is Open.

At the moment everything is closed, all the external contacts from the sensors go to the opposite condition which is NO. Internally the assigned contacts are NO hence, the timer is not active.

Since all the input contacts are in parallel connection, only one path or closed contact is needed to activate the timer. This last one waits for two seconds before activating the alarm.

In the second case:

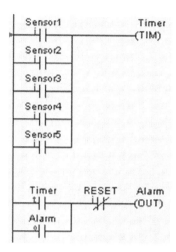

The sensors in normal condition are NO, it means that once the doors and windows are closed, they will close the contact and send an active signal to the PLC inputs. Since the internal contact we have assigned to the inputs are NC, then they will open in the ladder logic.

All the NC contacts related to the inputs are connected in series so, all of them must be closed in order to energize the auxiliary relay named Relay1. You can notice that a NC of Relay1 is also used on the second rung . It will remain open as long as the Relay1 is energized and this is only possible if all the safety of the house is OK and nobody wants to open any door or window.

When someone breaks into the house there is a delay of two second before activating the alarm. We are assuming here that the burglar doesn't know that there is an alarm system, because if he knows and manage to close the door or the window before two seconds, the alarm won't sound.

Simulation with the PLC trainer.

Here we assume that the RESET pushbutton is assigned to IN1. The red pushbutton on the trainer has a NO contact but that assigned contact in the ladder logic is a NC contact. This means that if we press the RESET button, the contact on the program will open.

The second pushbutton on the trainer has also a NO contact The four switches must be on the Off position or as a NO contact as well.

Press the second button or toggle any switch in less than two seconds and nothing will happen. If you last more than two seconds in this action then the output1, which is driving the alarm will become active.

You could also try to toggle two switches simultaneously and the effect will be the same since only one sensor is required to activate the alarm.

In order to reset the alarm just press the RESET button. If you don't do it, the alarm will keep active.

Notes

4. Counting of people entering into a place.

There is a sensor beam that a person will interrupt at the moment someone is entering. For every 10 persons, an audible alarm must sound just for two seconds, notifying that there is a group of ten people in the room. The same alarm will sound different if someone stays at the entrance, interrupting the beam for more than three seconds.

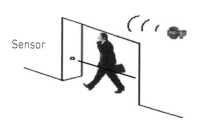

Sensor

Elements used:

Inputs:

> **Sensor:** Sensor NO, detect persons that cross the door.

Outputs:

> **Alarm:** Siren that sounds when 10 persons have entered.

Internal Relays:

Relay1: Retains the signal coming from the counter.

Timers:

Timer: Time to wait before activating the Alarm.

2sec-Timer: Time to sound the alarm when 10 persons have passed.

3sec-Timer: Time to force a person to move from the door when he is interrupting the beam of the sensor.

Counters:

N_ofPerson: Counts the number of persons (up to ten).

Sequencers:

None.

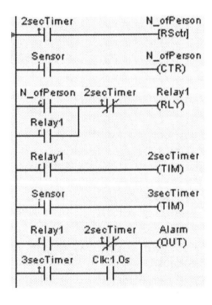

Program Operation.

The only input we have here is the Sensor that detects when the person crosses the door. You can notice that this contact is used for two purposes: to deliver the pulses to the counter that counts the Number of Persons and to drive or enable a 3 seconds timer.

You can observe that the first rung is similar to the one with Start and Stop button and the interlocking relay, which is a classical connection. The Start button has been replaced by the sensor.

Even when the person has passed across the door, the contact from Relay1 is retaining its own output. The reason is that in the next rung a timer of 2 seconds starts counting and then deactivates Relay1 again. This short time helps to avoid false triggering for the counter.

If for some reason the sensor remains active because the person is standing at the door, the 3 second timer starts counting. After the time runs out then the 3secTimer contact closes, enabling the automatic contact (special bit) Clk: 1:0s to keep on a continuous cycle (ON-OFF) . The final action is that the Alarm will be cycling between active for half a second and inactive for half a second, during the time the person remains hindering the sensor.

Simulation with the PLC Trainer.

If the Sensor is connected to IN1 on the trainer you can press the pushbutton to simulate the person passing across. You can check that on the Simulator screen, the two seconds counting on a 2secTimer. The counter starts from 10 and decreases once per every time you press the button, which is the same as having a person passing across the door. When it reaches zero it closes its contact and Relay1 contact performs the retention. No matter if you press the button and remove your finger very fast, the Alarm will sound for two seconds: The Output1 will remain two seconds ON , closing the its related contact.

You can keep your finger, on the pushbutton, for more than three seconds. When this happens, the contact from the timer 3secTimer closes and then the Alarm becomes active. The sound is intermittent as you can see from the LED1 on top of the physical relay which is changing from On to Off and viceversa, during three seconds.

Notes

5. Automatic Ice Cream Dispensing System.

A cone factory has a cream dispenser controlled by a valve that opens for two seconds when there is a Cone located exactly below the dispenser's output. The cones are transported by a conveyor belt. After pressing a Start button down, the belt starts moving until a cone is detected by a sensor, or, below the dispenser. By pressing down another button named Stop the whole process is suspended.

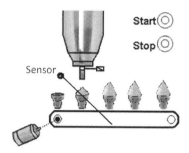

Elements used:

Inputs:

> **Start:** Pushbutton NO, to initiate the operation.
> **Stop:** Pushbutton NO, to stop the machine at any time.
> **Sensor:** Can be any type of sensor as long as it detects the piece.
> **Cone_Senso:** Sensor NO, detect when a cone is just below the cream dispenser.

Outputs:

> **Conveyor_M:** Drives the motor that is coupled to the conveyor.
> **Cream_Valv:** Drives the valve that dispenses the cream on the cone.

Internal Relays:

> **Relay1:** Retains the signal coming from the counter.

Timers:

> **2sec-Timer:** Time to dispense cream.

Counters:

    None.

Sequencers:

    None.

Program Operation.

Here we used the typical configuration with Start, Stop and interlocking relay to enable the machine operation.

You can notice that a contact from Relay has been used on the following two rungs, this means that the whole operation is disabled by forcing this contact to its normal condition , which is open. Once the pushbutton of Start is pressed down, this contact will be closed and the operation of the machine will rely on the rest of the ladder elements of the other rungs.

The sensor used to detect the cone is called Cone_Senso. If there is no cone the conveyor will move until a cone is detected. This can be said because if there is no cone, the contact from Cone_Senso will be closed , then the output that drives the conveyor is enabled (Conveyor_M).

On the other hand, in the last rung ,if there is a cone then the Cone_Senso will close and will energize the output that handles the Cream dispenser valve, named Cream_Valv. At the same time, with the same contact, a two seconds timer is energized. This is the time just required to dose the cream on the cone.

The contact from the timer 2sec_Timer is in parallel with the normally closed sensor contact as you can see it in the second rung. When there is a cone, this contact must open so, the timer contact will be the one in enabling the motor just enough time to remove the cone in front of the sensor. The band will move until a new cone is detected.

Simulation with the PLC trainer.

To start the process you can push the button on IN1. To stop the process at any time just push the second button, which is connected to IN2.

Both actions are either energizing or de-energizing the auxiliary relay named Relay.

For practical purposes you can assign the cone sensor to the toggle switch on IN3. Make sure that the corresponding light on the PLC is off before starting the simulation. This means that there is no cone and the Output1 will go ON, indicating that the conveyor belt is moving.

Change the switch from Off to On. The Output1 will turn off for two seconds and the Output2 will turn on for the same amount of time, which opens the valve that dispenses the cream. After the time has passed out the Output2 will turn off and Output1 will turn on again. This action takes place with the objective of advancing the conveyor until the detection of the new cone so, you can toggle the switch the off position again. The system is now in the original condition.

You can turn the switch to ON and the dispensing of creams starts again.

No matter what part of the process the machine is executing, if you press Stop, all outputs will go off and to run the machine you will have to start the machine again.

Notes

6. Level in liquids with turbulence.

A floating ball is coupled, through a long bar, to a switch. It´s detecting the moment the tank is losing certain level and sounding an alarm. The problem is that when some liquid is removed from the tank, by manually opening a valve, a turbulence is generated, affecting the activation of the switch. How can we prevent false triggering for the alarm?

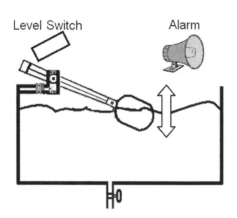

Elements used:

Inputs:

> **LevelSW:** Sensor NO, to detect directly or indirectly the movement of the arm of the pivoting float.

Outputs:

> **Alarm:** Drives an audible alarm

Internal Relays:

> None.

Timers:

2sec-Delay: Time to wait before activating the alarm.

Counters:

None.

Sequencers:

None.

Program operation.

This program is a very common solution to a sensor that has no stable detection.

In normal conditions , when the level is out of the required limit, then the level switch named LevelSW is detecting the float arm's position. The associated contact will open.

If the tank starts losing liquid level, then the lever will start rotating an angle until the sensor won't be able to detect it. It the non-detection persists for more than two seconds, the timer will run out closing its contact and then , it will activate the alarm.

If the sensor stops detecting the arm for less than two seconds and detects again, the timer will be de-energized and will have to start again.

For better operation you can move the sensor, change the shape of the arm detection area or change the timer's preset value.

Simulation with the PLC trainer.

The level sensor will be simulated by the pushbutton on IN1, but you can easily make the changes to the program to assign it to any of the switches.

After power up the PLC trainer, there will be a delay of two seconds and then the alarm, connected to the output1, will turn on.

Press and hold the button. The alarm will turn off. This would be the normal operation since the sensor would be always detecting the arm of the float. Remove your finger quickly and put it back, if this happens in less than two seconds the alarm will not get active.

Notes

7. Controlling Minimum and Maximum in water levels.

A water tank has two level sensors to detect Maximum and Minimum levels. There is an electric valve to fill the tank with water and there is a manual valve that is opened to feed another process. An automatic system is required to keep the water level under the limits. Once the low level is reached the tank must be filled up to the maximum. If after some time the maximum level has not been reached, the system must activate an alarm.

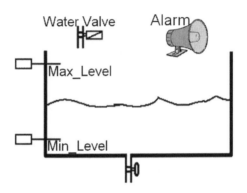

Elements used:

Inputs:

Max_Level: NO, detects the maximum of water allowed.

Min_Level: NO, detects the minimum of water allowed.

Outputs:

WaterValve: NC, allows to add water to the tank.

Alarm: Audible alarm to notify a water level problem.

Internal Relays:

> None.

Timers:

> **Delay:** Time to wait before activating the Alarm.

Counters:

> None.

Sequencers:

> None.

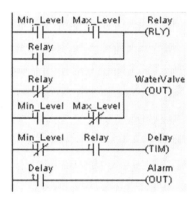

Program Operation.

For this program there is no need to use any pushbutton to start or stop the process. This is a continuous control and it´s worthless to increase the complexity for the ladder program.

When the program is started, we assume that the tank must be filled with water so, the first rung is checking when both level sensors (Min_Level and Max_Level) are detecting water at the same time. This means that the tank is ready to start the process.

You can notice that if the auxiliary relay named RELAY is not active, on the second rung, one of its NC contacts will energize the output WaterValve to make sure that water is entering to the tank to correct the levels.

If the water valve continues supplying water to the tank there will be a moment in which both level sensors will close its contacts, allowing the auxiliary relay to energize. If this happens, the NO contact in parallel, just on the first rung, will close. This action will keep Relay energized. The NC on the second rung will open, ending the water supply, since the WaterValve will de-energize.

You can notice the parallel branch formed by a series of a NO contact from Min_Level and a NC contact from Max_Level, will be the one on determining when the water valve will energize again since the NC of Relay is now open.

You can interpret this logic as saying: while the tank is having a minimum level (Min_Level will remain closed) and the level of maximum has not been reached (Max_Level is normally closed) the water valve must be energized.

The third rung is in charge of detecting when the tank is almost empty and below the minimum required level. Since the Min_Level contact is NC, then the timer called Delay will activate and count time. If the minimum level is not reached at certain time defined by the Delay´s Preset Value, the output named Alarm will get active.

Observe that this only works if the contact from Relay is closed and this only happens if the tank starts will a full water level.

Simulation with the PLC trainer.

The Output1 will start On. This means that the water valve is supplying water to the tank.

Press the pushbutton on the right (IN2) of the PLC trainer. This is simulating that the minimum level has been reached. You must keep pushing the pushbutton of IN2 and then press the red button on the left (IN1) which is assigned to the maximum level sensor. When both levels have been reached (both inputs are ON) then the output1 must turn off, meaning that the water supply must stop, sincenow the tank is full.

The initial conditions for the tank have been met and now water can be extracted manually. If the water level falls below the maximum level, just stop pressing the red button of IN1. Some water must be added to the tank so the Water Valve connected to Out1 will energize.

If the water is continuously drained, even while water is added, then the water will go below the minimum level. To simulate this just stop pressing IN2.

If you wait for some time (the Delay's preset value), meaning that the minimum level has not been reached, the output 2 will energize and the Alarm sound.

Try pressing the red button of IN2 at intervals. If the interval time is less than the delay, the alarm will never sound.

Notes

---

8. Printing machine.

The printed machine has a Printer arm, which is driven by a pneumatic piston. It´s in charge of stamping a logo on a paper sheet which is moving on a conveyor belt. There is a sensor to detect that the sheet is in place so the motor stops while the printing process takes place. Another sensor verifies that the Printer arm is up or down, to prevent that the belt moves while the arm is stamping. The motor coupled to the belt must run forward until the sensor is exactly below the printer arm.

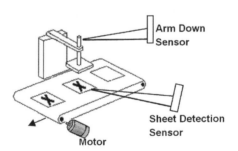

Elements used:

Inputs:

> **Start:** Pushbutton NO, to initiate the operation.
> **Stop:** Pushbutton NO, to stop the machine at any time.
> **SheetDetec:** Can be any type of sensor as long as it detects the piece.
> **ArmDown:** Sensor NO, detects when the Arm is printing.

Outputs:

> **Motor:** Turns the motor on.
> **PrinterArm:** Energizes a valve that sends the Printer Arm down.

Internal Relays:

> **Relay1:** Enables the system to work after Start is pressed.
> **Relay2:** Helps to generate a pulse to return the PrinterArm back.

Timers:

> **1SecDelay:**Time to release PrinterArm when. finished printing
> **2SecDelay:** Time to print.

Counters:

> None.

Sequencers:

> None.

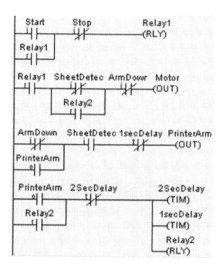

Program Operation.

After pressing pushbutton Start, the auxiliary relay named Relay1 will get energized closing its contact, just below Start. This will keep it energized until Stop is pressed and removes the power on it.

On the second rung, the contact from Relay is closed, enabling the rest of the remaining ladder elements. The initial condition for the contacts of SheetDetect and ArmDown is normally close so, the output Motor will turn on. This means that the Motor will start running and will continue so until a paper sheet is detected by the sensor, connected to SheetDetec input.

On the third rung, the only contact that prevents the activation of the output named PrinterArm, is precisely SheetDetec. Since this contact is closed now, assuming that there are sheets on the belt, the printer arm will start going down.

Pay special attention to the fact that the sensor used to detect the position of the Printer Arm is located, for practical purposes, on top of the machine so, while the Arm is up, the sensor of ArmDown is detecting but not sending any signal to the plc input.

At the moment the ArmDown sensor stops sensing the Arm it will change the status of all ist contacts: all those that were closed will open and viceversa. This is the reason for us to use a NO contact from the PrinterARm output just in parallel with the ArmDown contact: to retain the output PrinterArm the necessary time to do the job, which is assumed to be one second.

On the third rung, the contact from the output PrinterArm turns on three elements on the right: two timers and one auxiliary relay. This last one is in charge of allowing the two timers to remain energized after the first second when the PrinterArm output is turned off, forcing the arm to go up again.

Simulation with the PLC trainer.

This is one very good example where the software simulation can be very tricky to use in the understanding of the program. Some signals are occurring simultaneously so it makes it difficult to visualize the current operation.

To start just push the first pushbutton which has been assigned to Start. The output 1 goes on , meaning that the motor of the band starts running. It will continue running until there is a sheet detected and since this sensor is assigned to IN3, just toggle the switch to the ON position (you will notice that it´s on because the green LED on the PLC is also on). This will start the printing process:the output2 turns on for one second and then turns off.

It means that the Printer Arm went down and then returns to its original position.

Try to synchronize the toggling of the output2 with you moving the lever of input 3 switch on and off. This is the sensor that detects that the printer is going down so , it will turn off the output1. Notice that this is the only way for turning it off.

You use the Arm Down sensor connected to IN4 to make sure the motor is not running when you are printing.

At any moment you press stop every output and internal relay goes off.

Notes

---

9. Animatronic head.

There is an animated head that has moving eyes and mouth. It can be used on stores to gain customers' attention. The talking head has the following sequential movements:

Blinks the eyes, move the mouth twice, waits two seconds, blinks the eyes, waits for 10 seconds and starts the cycle again.

For practical purposes we don't have to worry about the components for the mechanisms. We can assume that they only require a simple activation or de-activation signal.

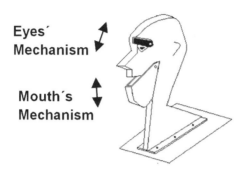

Elements used:

Inputs:

> None.

Outputs:

> **Eyes:** Activates the Eyes´ mechanism.
> **Mouth:** Activates the Mouth´s mechanism.

Internal Relays:

> None.

Timers:

> **1secdelay:**Time to re-start the cycle.
> **2secdelay:**Timefortheaction.EitherEyesorMouth.
> **10secdelay:** Time to wait before to re-start the
>   cycle.

Counters:

None.

Sequencers:

None.

```
       Timer1   Timer2                 Eyes
    ─────┤ ├──────┤/├─────────────────(OUT)

     2secdelay Timer2  Clk:1.0s        Mouth
    ────┤/├──────┤ ├──────┤ ├─────────(OUT)

     10secdelay                        Timer1
    ────┤/├───────────────────────────(TIM)

       Timer1                          Timer2
    ─────┤ ├──────────────────────────(TIM)

       Timer2                          2secdelay
    ─────┤ ├──────────────────────────(TIM)

     2secdelay                         10secdelay
    ─────┤ ├──────────────────────────(TIM)
```

Program Operation

Since this is an automatic and continuous operation, there is no need to use inputs.

The first two rungs contain the ladder logic to control the only two outputs that we have here:

Eyes and Mouth.

For the following rungs, you can notice that all the timers are connected in "cascade". This means that every timer, once it´s elapsed, will enable the next one.

The last timer 10secdelay is in charge of re-starting the cycle. You can see that a normally closed contact is used to enable Timer1 which is the one that starts the cycle.

On the second rung, when both timers have the contact normally closed then the activation of the mouth depends exclusively on Clk:1.0s. This contact, as you probably know, is automatically opening and closing at a speed of once per second (1Hz).

Simulation with the PLC trainer

When the PLC trainer is powered up and assuming that you have downloaded the program, the cycle will start automatically. There is no need to press any button, according to the design requirements.

After some time, defined by the Timer1's preset value , the output 1 will turn on and hence the eyes will move to one position. Some time later, when Timer2 passes out, the eyes will go to the original position. You will see that output1 turns off.

The output 2 , the mouth, starts to change from on to off and vice versa. It´s expected to have to move the mouth two times during two seconds (once per second).

There is a 10 seconds delay and the outputs start to turn on in the same sequence as described before.

Notes

10.A. Talking machines (First solution)

Most of the machines you know are probably using lights to notify a failure or any process condition. Using the Voice Module (http://www.lt-automation.com/VoiceModule.htm), you can add voice to any machine in a very innovative way. Previously, program the module with three alarm messages (IE. "High temperature in the oven", "The pressure is below the limits" and "Please close the door before operating the machine").

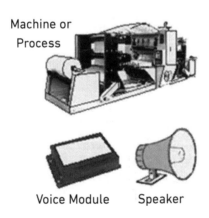

Machine or Process

Voice Module   Speaker

Write a program to play all three messages sequentially. Every message has three seconds duration (for demonstration purposes). There is no need to wait among messages.

Elements used:

Inputs:

> **Start:** Pushbutton NO, to initiate the operation.
> **Stop:** Pushbutton NO, to stop the machine at any time.

Outputs:

> **Out1:** Play message 1.
> **Out2:** Play message 2.
> **Out3:** Play message 3.

Internal Relays:

> **Relay1:** To retain the running operation once Start is pressed.

Timers:

> **Timer1:**Time to put the outputs combination for 1.
> **Timer2:** Time to put the outputs combination for 2.
> **Timer3:** Time to put the outputs combination for 3.
> **Timer4:** Time re-start the cycle.

Counters:

> None.

Sequencers:

> None.

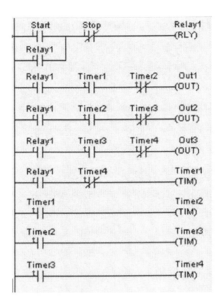

Program Operation.

You can notice that we use Start, Stop and Relay1 to configure a common ladder logic which allows us to enable the whole operation.

The logic is directed to create a combination of the outputs in such a way that only one output is active at a given time.

The voice module has 8 inputs that can be programmed to play the messages in several ways:

1. **Direct:** Every input play one message.
2. **Binary:** The binary combination of the inputs select the message to play.
3. **Serial communication:** A code sent to the communication port of the module will play a given message.

The one we will use for a program is the first one: Direct.

Once you press Start the Relay1 becomes active and remain energized. Contacts from Relay1 are used on subsequent rungs to enable the logic.

You can notice that Timer1 is the first element in getting active (rung number 5), and then there is a cascade of activations for the rest of the timers. Once Timer4 has passed out, the cycle re-starts again.

You can observe that only a contact from Timer4 is used to de-energize Timer1 and not the rest of the timers.

The operation is automatic and continuous and only stops at the moment the pushbutton Stop is pressed.

Simulation with the PLC trainer.

If the program has been downloaded, you only need to press the red pushbutton on the left that has been assigned to Start. Some time later (Timer1) Output1 will turn on and will remain on for the time that Timer2 has been programmed.

The same happens for Output2 and Output3 on the trainer.

Since every output of the PLC trainer is driving a relay, the contact from each of them can be used to provide the signal to the voice module. To perform the right wiring you can refer to the manual of the module itself.

At any time, if you press the Stop button (the other red button) the sequence of outputs will stop immediately.

Notes

10.B. Talking machines (Second solution).

This is another program version. It uses some other instructions like the master control (with ILock and ILoff), and a 4 step sequencer.

Elements used:

Inputs:

Start: Pushbutton NO, to initiate the operation.
Stop: Pushbutton NO, to stop the machine at any time.

Outputs:

Out1: Play message1.
Out2: Play message2.
Out3: Play message3.

Internal Relays:

Relay1: To retain the running operation once Start is pressed.

Relay2: To re-start the counting pulse (clock).

Timers:

Timer1: Handles the Pulse duration to provides the counting (clock).

Counters:

Seq1: Associated with the Sequencer.

Sequencers:

Seq1: A sequencer of 4 steps to generate a binary combination with the outputs.

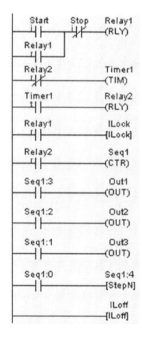

Program Operation.

To enable the system to work we need to activate Relay1 and this can be done by simply pressing Start. The relay is maintained because of its contact in parallel with Start.

Relay1 allows that all the logic between the instructions of ILock and ILoff can work. This includes the activation of the outputs required to play the messages on the module.

Here we use a sequencer of 4 steps that is associated to a Counter precisely denoted Seq1.

A combination of the step contacts of the sequencer is required to activate the specific output at a given time.

Since the counter of the sequencer requires pulses to count and follow the sequence steps, then we use a contact from Relay2 as the input for the counter Seq1.

Relay2 and Timer1 are connected as a continuous pulse generator that opens and closes the Relay2's contact.

When the sequencer reaches zero the contact Seq1:0 closes and forces the sequencer to jump to the step 4, through the command StepN. It's exactly the same as programming the set value of the counter with the number 4.

Simulation with the PLC trainer.

The simulation is exactly the same as the previous version of the ladder program.

You press Start and the PLC outputs will turn on in sequence from Out1 up to Out3.

The speed of this sequence is determined by the preset value you have selected for Timer1 .

To stop the program at any time, just press Start.

Notes

11. Controlling a DC motor. Forward/Reverse

There is a DC motor that must be controlled by a ladder program. To make the application really interesting, write a program with Manual/Automatic operation. If Manual is selected there is a switch selector (Forward/Reverse) to determine the direction of rotation of the motor. In Automatic the motor keeps continuosly running between forward for 3 seconds and reverse for a different time of 5 seconds.

Counter clockwise

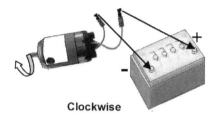

Clockwise

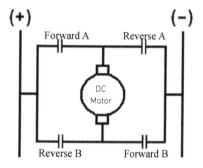

Elements used:

Inputs:

> **Start:** Pushbutton NO, to initiate the operation.
> **Stop:** Pushbutton NO, to stop the machine at any time.
> **Man_Auto:** Selector switch. One position is for Manual operation and the other one is for Automatic operation.
> **Forw_Rev:** Selector switch. One position is for

Forward motor rotation and the other is for Reverse rotation.

Outputs:

ForwardA: energizes output1 that drives Forward direction.
ForwardB: energizes output2 that drives Forward direction.
ReverseA: energizes output3 that drives Forward direction.
ReverseB: energizes output3 that drives Forward direction.

Internal Relays:

Relay: to retain the running operation once Start is pressed.

Timers:

3secTimer:Time for forward motor rotation.
5secTimer: Time for reverse motor rotation.

Counters:

None.

Sequencers:

None.

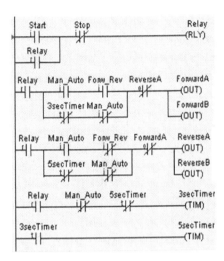

Program Operation

In general terms, you need to change the polarity of the voltage, applied to its terminals, to change the sense of rotation of any DC motor. For this purpose, we have used four outputs: two for forward and two for reverse rotation. Somehow this is a waste of outputs since you could use a single output driving a relay with two contacts for each rotation sense. The reason to use four outputs is because you don't need to use extra devices when using our PLC trainers.

When you press Start the interior contact Relay will energize, generating a voltage support path for itself. In the subsequent rungs there is a contact from Relay, meaning that the rest of the program depends on the enabling of this particular interior relay (Relay).

The next two rungs look similar, but there are some differences.

The contact Forw_Rev appears NO on the second rung and NC on the third rung, meaning that only one of the rungs will work at any given time: only the specific outputs related with the sense of rotation that has been selected will get active.

In order to "block" a possible simultaneous operation of having forward and reverse outputs at the same time, a contact from one of the outputs is added at the end of the rung in charge of the contrary rotation.

The last two rungs are forming a clock signal allowing the DC motor to rung 3 seconds in the forward sense and 5 seconds in the reverse sense.

In addition to Start and Stop buttons, there are two selector switches named Man_Auto, to allow Manual operation (one sense of rotation) and Automatic operation (continuous changes of rotation sense) and the other selector is Forward or Reverse that only works in manual operation.

The change of sense of rotation, when Automatic mode is selected, is determined by the closing of the timer´s contacts.

Simulation with the PLC trainer

Move the first and the second toggle switches on the trainers to the ON position, this is needed if we want to start with manual operation and with forward rotation selected.

Press Start, which is the first pushbutton, to start the motor. You notice that the first two outputs Out1 and Out2 , corresponding to ForwardA and ForwardB, will turn on. You can wait as long as you want and they will remain active always.

If you use the software simulator you notice that the timers are working but they don't affect the operation since the manual option is selected.

To change to Reverse rotation move the second toggle switch to the Off position. The outputs Out3 and Out4 will turn On (ReverseA and ReverseB) while the first two turn off. This means that the motor, in manual mode, is rotating on a contrary sense than before.

You can change the first toggle switch (Manual or Automatic selector) to the Off position, meaning that you have selected the automatic operation. Through the online monitoring software, you can realize that now the timers are the ones determining the sense of rotation.

At any time that you press Stop, the motor stops since the outputs turns off and the program is forced to reinitiate because the internal relay named Relay is de-activated.

Notes

12. Starting a tri-phase AC motor with a PLC (Star-Delta connection).

The starting current of a tri-phase motor must be limited in order to have a proper operation. You need to change the motor windings connections from the "way" to "delta" connection (also known as Y-Delta). To do this, you need two contactors and a 3 second timer. Write a program to make a soft start of any tri-phase motor through the Y-Delta connection.

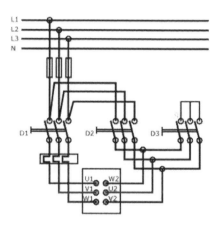

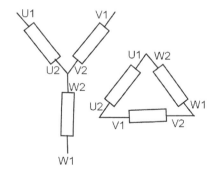

Elements used:

Inputs:

> **Start**: Pushbutton NO, to turn the motor On.
> **Stop**: Pushbutton NO, to stop the motor at any time.

Outputs:

> **Y_Connecti**: energizes output1 that connects the windings in Y.
> **Delta_Conn**: energizes output2 that connects the windings in Delta.

Internal Relays:

> **Relay**: to retain the running operation once Start is pressed.

Timers:

> **Timer**: Time to wait before changing from Y to Delta connection.

Counters:

> None.

Sequencers:

> None.

```
     Start      Stop              Relay1
    ─┤ ├───────┤/├──────────────(RLY)
     Relay1
    ─┤ ├─┘


     Relay1                       Timer
    ─┤ ├────────────────────────(TIM)


     Relay1  Delta_Conn Timer  Y_Connecti
    ─┤ ├────────┤/├──────┤/├────(OUT)


     Relay1    Timer            Delta_Conn
    ─┤ ├───────┤ ├─────────────(OUT)
```

Program Operation.

Start, Stop and Relay1 interact together to create a way to enable the electrical circuit to work by energizing and retaining the auxiliary relay named Relay1.

Once the Relay1 is energized, by pressing the pushbutton Start, all its contacts close allowing the next three rungs to work.

The Timer is now energized and starts counting time on the second rung.

The third rung makes the Y connection immediately by activating the Y-Connecti output. This is a normal requirement since every motor must start on Star (Y) connection. When the timer has counted three seconds then it will close its contact on the last rung and, since the rest of the contacts are already closed, the Delta_Conn will get active.

To prevent that both outputs are active at the same time, a contact from the output Delta_Conn is put on the third rung, meaning that if the Y_Connecti output can't be energized while Delta_Conn is energized.

At any time, you can press Stop and both outputs will turn off because Relay1 is no active anymore and all its related contacts will open.

Simulation with the PLC trainer.

As soon as you press the first red pushbutton (assigned to Start), the output1 will turn on. This means that the Y connection has taken place.

If there is a motor it will start running. Three seconds later the Output1 will turn off, while the Output2 will turn on, meaning that the motor windings are in Delta connection.

The main reason to change from Y to Delta is to have the maximum torque at the moment the motor is starting, when it has overpassed the initial inertia, the current required can be reduced by changing the connection of the windings.

Output1 and Output2 from the PLC trainer are used to drive two different relays or contactors with three high current contacts that are connected according to the electrical diagram shown before.

In addition to the ladder logic interlocking we use, to prevent that only one output is active at a time, a physical interlocking is made through some auxiliary contacts which are available on the contactors.

<u>Notes</u>

13. Controlling a tri-phase AC motor. Forward/ Reverse

Forward and reverse rotation operation on a tri-phase AC motor requires using of contactors to switch just two of the input voltage lines. Write a program that makes this automatic operation having the motor rotating 10 seconds forward and 8 seconds in reverse mode, on a continuous way.

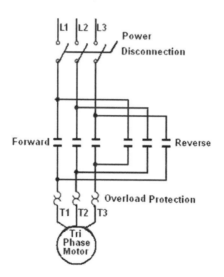

Elements used:

Inputs:

> **Start**: Pushbutton NO, to initiate the operation of the motor.
> **Stop**: Pushbutton NO, to stop the motor at any time.

Outputs:

> **Forward**: energizes output that drives forward direction.
> **Reverse**: energizes output that drives reverse direction.

Internal Relays:

> *Relay*: to retain the running operation once Start is pressed.

Timers:

> **10secTimer**:Time for forward rotation of the motor.
> **8secTimer**:Time for Reverse rotation of the motor.

Counters:

> None.

Sequencers:

> None.

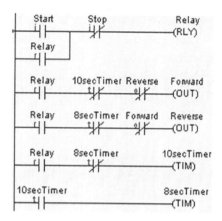

Program Operation.

Closing the contact from Start, the Relay will energize and retain itself. The contact in parallel with Start will keep it energized even if the contact Start is not closed anymore.

On the second rung, all the contacts are closed since Relay is closed after pressing Start, then the output Forward will get active. This means that the motor is energized through any device (a Contactor or Relay) that applies power to its three terminals.

The NC contact from 10secTimer will be the one in charge of stopping the forward movement of the motor. This contact depends on the logic on rung number 4, since the 10secTimer will start right after you press Start and the Relay contact is closed.

While the motor is running forward or the output Forward is active, the contact on the rung 3, that is a NC contact, will be open. This prevents that both outputs or contactors be active at the same time.

When the motor stops running in forward direction, after the 10 seconds (from the 10secTimer) have run out, the NC contact from Forward closes energizing the Reverse output. At the same time, because the 10secTimer has elapsed, its contact is closed and the 8secTimer of the last rung is enabled.

Once the 8secTimer stops counting its time, it simultaneously de-energize the output Reverse, enables the Forward output and the 10secTimer gets active.

If the contact from Stop is opened at any time, all the operation ends and all the outputs are turned off. It will require that you press Start to re-initiate the process again.

Simulation with the PLC trainer.

When you press Start the output1 will turn on, meaning that the motor is starting to run in the Forward direction. The contact between R1A and R1B of the PLC trainer, will be driving the contactor in charge of the forward direction.

After 10 seconds, the output1 will be turned off and then the output2 will turn on. The contact between R2A and R2B will energize the contactor driving the reverse direction.

You can notice that the connection is about the same with the only difference that two of the power input lines have been swapped.

The output2 will be active for 8 seconds and then it will turn off. The cycle will start by turning on Output1 again.

Pressing stop at any time, no matter which part of the cycle the control is, the whole process will end.

In practical terms, it's much better to wait for some time until the motor has lost most its inertial motion before trying to run it in the opposite direction.

Notes

14. Controlling a Stepper motor through a driver.

One very common application is related to using PLCs to command Stepper motors, through a simple ladder logic PLC program.

Write a program that sends 50 pulses in one direction and 30 pulses on the other direction. Assume that you have a Stepper motor driver with two main inputs: Direction and Pulses.

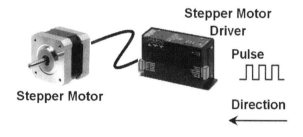

Stepper Motor Driver

Stepper Motor

Pulse

Direction

Elements used:

Inputs:

> **Start**: Pushbutton NO, to initiate the operation of the motor.
> **Stop**: Pushbutton NO, to stop the motor at any time.

Outputs:

> **Direction**: Toggles when every counting is completed. If it was on then turns off and vice versa.
> **Pulses**: Output that provides the pulses to the drive.

Internal Relays:

> **Relay**: to retain the running operation once Start is pressed.
> **RelayAux1 and RelayAux2:** help with the logic to know the last state from the output Direction.

Timers:

> None.

Counters:

> **30_pulses**: Counts for 30 pulses.
> **50_pulses**: Counts for 50 pulses.

Sequencers:

> None.

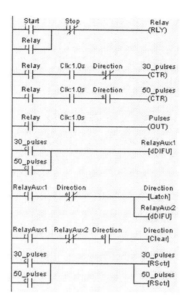

Program Operation.

Start, Stop and Relay interact together to create a way to enable the system to work. When the contact from Start is closed then the auxiliary relay named Relay will energize and close its contacts. The contact it has below Start will help it to remain energized until the contact from Stop opens.

The second rung shows an automatic clock of 1 second named Clk:1.0s, in series with a contact from Relay. This special bit opens and closes continuously at a frequency of one time per second. This clock signal is used for two purposes: sending the counting pulses to the 30 and 50 pulse counters and producing the activation and deactivation of Output1, which in turn sends the external pulse to the driver.

The third rung uses the counters' contact (30 and 50) in parallel connection to energize an auxiliary relay named RelayAux1 through a special command dDIFU. This command activates the output only at the transition from open to close according to the logic, previous to the output.

The rungs number 4 and 5 are in charge of "remembering" the last status for the output named Direction and changing it to the opposite status, every time a counter reaches its maximum counting (counter's preset value). The final result is that the output Direction toggles between activation or deactivation. This output determines the sense of rotation the Stepper will take.

The command "Latch" is used to turn on (and leave on) the output named Direction and the other command "Clear" will turn off the same output.

The last rung resets the counters every time they reach the set value and leaves them ready for the next counting action.

The process can be reset at any time by pressing Stop.

Simulation with the PLC trainer.

Pressing Start will initiate the operation. The output2 will begin to cycle continuously between on and off, providing the pulses to the stepper motor drive. After 30 pulses on output2 the output Direction will turn on and will remain on for 50 pulses.

In general terms you will see Direction off for 30 pulses and ON for 50 pulses. The pulses are represented by the connection and disconnection of Output1.

Hooking up the drive to the trainer is a matter of using two contacts available at the output connector. For proper connections check the manual that accompanies the drive.

You can notice that contacts of these outputs are included in the logic of the last two rungs.

Notes

WIRING A PLC AND A PLC TRAINER

14

How to make connections to the inputs and outputs of the PLCs. If you have a PLC trainer you can practice with real devices.

14. Wiring devices to the Inputs.

From previous chapters you have learnt that PLC inputs can be classified in two:

DC or AC.

14.1 DC Inputs.

The DC inputs are considered as "low voltage" inputs that can also be subdivided in other two:

NPN or PNP.

This kind of classification is based on the transistor type used for the electronics on the input signal.

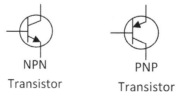

NPN
Transistor

PNP
Transistor

The small arrow shows the sense of the current. You need to be very careful in knowing the type of PLC input.

To connect the inputs you must do the following:

NPN inputs

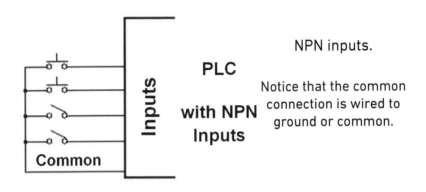

NPN inputs.

Notice that the common connection is wired to ground or common.

PNP inputs

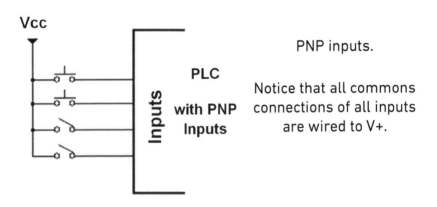

PNP inputs.

Notice that all commons connections of all inputs are wired to V+.

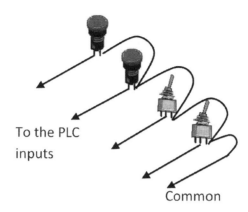

To the PLC
inputs

Common

From the graphic you can see that if we want to connect push Buttons and switches, only for explanation purposes, we connect a common line to one end of each of these devices (which normally are two terminal devices) and the other end goes directly to the PLC input.

The common terminal must be connected either to Vcc in case of PLCs with PNP inputs or to ground 0V, or the common of the PLC if the PLC receives NPN inputs.

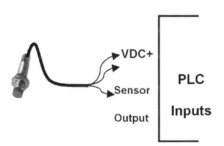

To connect the electronic sensor either inductive or capacitive type, you can have many ways.

The DC powered electronic sensors (inductive and capacitive) don't require a common wire, since they are already connected to the power supply, which in most of the cases is the same for the PLC.

Inductive and capacitive sensors are specified by a number of parameters like size, sensing distance, connection, etc.

It's very important to know the kind of output they have: NPN or PNP.

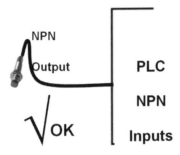

The sensor output and the PLC input must be the same:

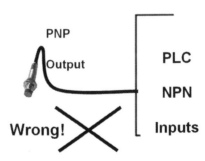

14.1.2 AC inputs.

In similar way like the DC inputs, the PLC with AC inputs have a particular way to be connected.

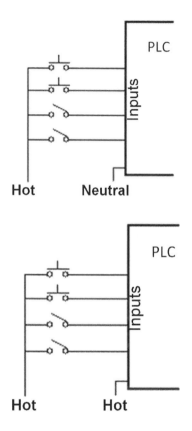

In general, the working voltages can be either 110VAC or 220VAC, The wiring requires a proper connection and care.

The above figures show the way you need to connect the devices' common. The PLCs with AC inputs have a connection point for the other AC voltage supply terminal.

14.2 Wiring devices to the Outputs.

Since there are three different types of in PLC outputs, there are also several ways to connect the outputs to the loads.

14.2.1 DC outputs.

The DC output, in most of the PLCs, has a low current capacity. This is why it´s not always possible to connect them directly to other devices. Many of them require a current level higher than the one the output can drive.

We can have the following cases:

1. PLC Output to PLC input

A PLC can send a signal to another PLC as long as the PLC output and the PLC input are compatible. This means the same type.

Example: PNP output to PNP input.

2. PLC to Electronic device

The PLC can send control signals to an electronic device, which in turn will drive another load (IE a motor). The electronic device must receive a signal compatible in both : type and level.

Some PLC outputs are capable of generating high speed outputs for duty cycle control.

3. PLC driving an Opto-coupler

In order to protect the PLC from the effects of high current devices, it's very common to have the PLC output driving a LED or an Opto-coupler.

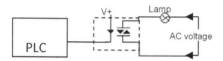

You can notice that the two circuits: PLC side and Load side are isolated by the opto-coupler. This means that the PLC turns on the LED and then it sends light as a command to close the other part of the semiconductor. This last one acts as a switch.

4. PLC driving and external transistor.

When there is a need to drive higher current signal, an external transistor with higher current capability is used. In this case, the PLC is connected to the terminal of the transistor (named Base). The transistor will act as a switch, connecting or disconnecting the load.

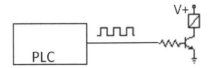

Some PLCs can produce Pulse Width Modulated outputs, when applied to a transistor that drives a coil from a valve, then it's possible to control (proportionally) the aperture of the mechanical valve.

5. PLC driving an external Relay

This is the most common application of using a DC PLC output. This is because of two important facts:

a) If the output requires to connect or disconnect a DC or AC higher current load then you need to use an external relay. This Relay can be Electromechanical or Solid State type.

b) If the connection or disconnection is going to be intensive, then the contacts' life from the relay will be affected with the time and at some point of your process then will fail. To facilitate the replacement of a new relay it's much better to have them externally connected. These relays can be Interface relays or Multi-contact heavy duty relays.

According to the PLC output, if it´s NPN or PNP you have a different way to wire the external relay.

The sinking connection is shown on previous page. The PLC has an NPN output and provides the path to ground for the relay coil.

The sourcing connection is shown above. The PLC has an PNP output and provides the power connection or the V+ to the relay coil.

14.2.2 AC outputs

The AC PLC outputs behave the same no matter if they are 110 or 220VAC.

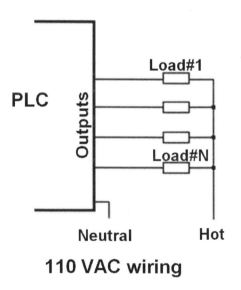

110 VAC wiring

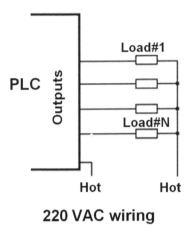

220 VAC wiring

Most of the cases the AC outputs are used to drive contactor or relay coils as loads, so the outputs have enough current capability to drive theses ones. Usually it's about 80 milliamps.

Since the loads are two terminal devices we have to make one common connection for each device, by connecting one of the terminals to a Phase (also known as Hot) wire. The other terminal of each device must be connected to a specific output in the PLC.

Working with AC is somehow risky. You should pay attention in handling AC lines and making sure that you are correctly wiring the system.

14.2.3 Relay Outputs.

As you probably know, we are talking about the physical and electromagnetic PLC relays, that can be soldered inside the PLC board when it´s a fixed architecture PLC or the relay that is in the Relay module when the PLC is modular or expandable.

Although manufacturers have developed smaller and more powerful relays, the relay contacts have some restriction on the amperage they can handle and the size they occupy inside the PLC board or module.

Again, if the relay output is going to be constantly connecting and disconnecting during the PLC operation, it´s much better to use an external relay.

The big advantage of using relay outputs is that they can be connected as a switch in either DC or AC circuits. The disadvantages are the speed of the device and the number of activations, which in terms of damages, make them difficult to repair.

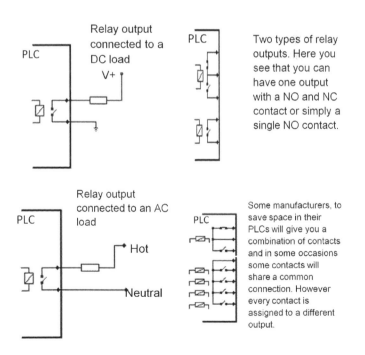

Relay output connected to a DC load

V+

Two types of relay outputs. Here you see that you can have one output with a NO and NC contact or simply a single NO contact.

Relay output connected to an AC load

Hot

Neutral

Some manufacturers, to save space in their PLCs will give you a combination of contacts and in some occasions some contacts will share a common connection. However every contact is assigned to a different output.

You need to be very careful in using the relay terminals and its contacts. If you exceed the amperage specification then you can damage the PLC output or the PLC itself.

14.3 Connecting devices to the PLC trainer.

14.3.1 Inputs in the PLC trainer.

You can make connections of real sensors to the PLC trainer.

If you have the PLC trainer,just take a look to the connectors with labels In1, In2, In3, .. In6.

Once you have identified the digital input connector, you are ready to hook up your devices.

A very important remark: the digital inputs, available on the connector, are also connected to the Pushbuttons and the switches on the PLC trainer. This means that you need to be very careful when connecting external devices because of a possibility of a short circuit.

A short circuit can produce damages to the device and the PLC trainer and it happens because it´s possible that the device you plan to connect is sending an V+ voltage to the PLC trainer, while any of the push buttons or the switches will send a Ground or zero volts, when they are activated.

To avoid any short circuit , make sure that inputs In3 up to In6 are off (the Green LEDs on the PLC are off). This is not needed for Input1 and Input 2 which are always off (unless you press the buttons).

The next figure shows the Plc trainer connector for DC inputs. You can notice that the first and the last two positions are reserved for power supply. These are OUTPUTS and not inputs, this means that they provide a voltage output, which come from the main power supply (or wall adapter).

The current capability of theses outputs is not that high and will only work with some common sensors that only require small currents.

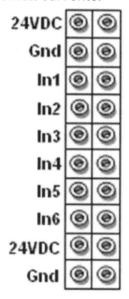

Remember that only devices with NPN outputs can be connected to the PLC trainer.

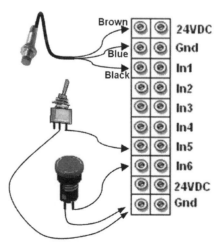

The figure shows the way to make the physical connections. We have moved the labels to the right to have a clearer view

As a general rule, the DC sensors follow a color code for the outputs so: Brown is positive, Blue is negative and Black is the color for the output.

Any other device, that doesn't require to be powered, like a switch or pushbutton, can be connected as shown on the figure. One terminal to GND or Zero volts and the other to the input.

14.3.2 Outputs in the PLC trainer.

If you are using the PLC trainer the right side connector will allow you to wire external devices to the outputs.

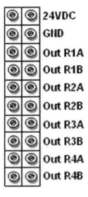

The PLC trainer has external relays, located on the small printed circuit board located at the right side of the PLC. The relays on the board receive the activation signal directly from the PLC outputs.

The contacts from each relay have been wired to the side connector as described on the next page figure:

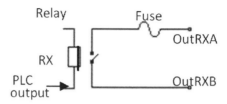

X is a relay number (1,2,3 and 4)

You notice that the contact terminals are denoted by RXA and RXB to differentiate them from the other relay contacts. For example OutR3A and Out R3B correspond to the Relay# 3 and are controlled by the PLC output #3.

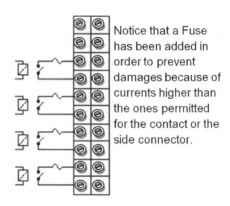

Notice that a Fuse has been added in order to prevent damages because of currents higher than the ones permitted for the contact or the side connector.

To connect any AC or DC load, you can implement any of the circuits shown on the next page. As you can see, both types of loads can be connected to any of the relays.

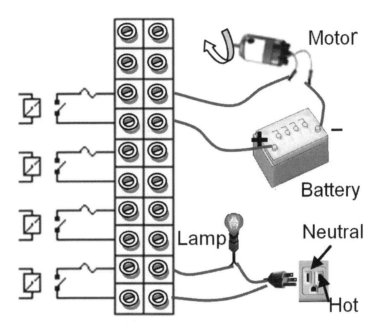

You can use the DC Power supply outputs available on the connector, but only when your load requires low current levels. It's better for you to use an external DC power supply like the battery shown on the above figure.

Safe AC connections.

A safe way to connect AC loads or devices to the PLC trainer can be implemented as follows:

1) Get an extension cord

2) Approximately at 10 inches from the AC outlet, cut one of the wires.

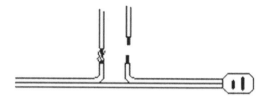

3) Attach extension cables, properly pealed on every end.

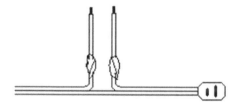

4) Cover the junction with insulating tape.

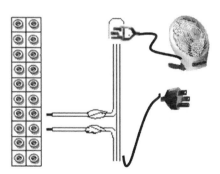

Now you have an easy and safe element to connect any AC device to the PLC trainer and make your applications with a more realistic feeling.

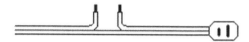

GLOSSARY

Definitions and terms commonly used in automation.

A

AC

Alternate Current. Power system where the current flow changes in polarity at a given frequency.

Active

For outputs or devices. It means connected or working.

Actuator

Element that will produce the final action like change of position, aperture or closing of gates , etc.

Address

It´s a number assigned to a device when having more than one (also known as network).

Analog

It´s referred to signals that can take any value in time. Opposite to Digital.

Animatronics

The technology of automation and robotics applied to puppets that seem to look alive.

Alarm

Notice or signal to notify a wrong operation.

Amp

Unit of Current. It's the result of dividing Voltage by Resistance.

Amperage

Term related with the amount of current flowing through a circuit.

Application technician

Person with technical background who develops automation projects.

Automation

The techniques and equipment used to obtain a better, more efficient and safer process operation.

B

Basic language

Programming language characterized by a simple and powerful set of instructions.

Board

Plate of fiberglass or Bakelite which contains electronic components.

Branch

In an electrical circuit, another possible path.

Bridge

Group of Diodes used to rectify the AC voltage and turn it into DC.

Connection between two electronic equipments with different communication protocol.

Building automation.

Automation technology specialized in buildings, houses, outlets,etc. Also known as Domotics.

Button

Device that acts an input of a PLC.

Buzzer

A sort of alarm with a characteristic sound produced by a vibration.

C

Cascade

In electronics it´s related to a device acting on another one. The location reflects the importance or priority.

Chassis

External box to contain a single or a group of boards.

Circuit

A combination of different elements like power supply, switches and loads, which has an operational purpose.

Cycle

For periodic electronic signals, the duration of the On and Off times or the duration of the whole signal that repeats periodically.

Current

Electron flow produced by a Voltage source applied to a Load.

Coil

In PLCs it´s an output that can be internal or external.

In electrical devices is a system of multiple wire turns around a ferrous core.

Controller

Device that, once it´s programmed, has autonomous capabilities to perform a control task.

Counter

Software element that allows the counting of objects.

Compact Flash

Device of electronic memory to store data that can be removed from the equipment.

Communication

Capability of an electronic device to send or receive data, usually through a communication port.

D

Data

Digital information handled by the main processor.

De-energized

Without energy. Inactive.

Delta

Type of connection of three loads, where the image looks like a Delta symbol or a triangle.

Detection

The status obtained when an object or event is sensed by the PLC.

Delay

Waiting time before performing a control action.

Differential

In PLCs it´s an action performed on one of the edges of the transition from connection to disconnection and viceversa.

Download

Action of sending the program from one equipment to another. For instance, the control program is downloaded from the PC to the PLC.

Domotics

It´s the automation and robotics applied to the home appliances.

Drive

A device with specific purpose of controlling a motor.

Duty Cycle

The relation between the on time and the total duration (Ton +Toff) of a signal.

E

Edit

In PLC programming, the state where you can write or make changes to a program.

EEPROM

Electrical Erasable and Programmable Read Only Memory. A modern technology in data memory that allows writing to the memory by electric means. The EEPROM is usually used to store the control program.

Electromagnetic

The resultant field produced by electrical current flowing through a wire.

Electro-valve

Valve that is opened or closed by electric means.

Electricity

All type of devices and circuits that involved flow of electrons or in general, electrical charges.

Emergency

Situation of danger that requires immediate attention.

Enable

Signal that allows operation of a device, a circuit, a rung, etc.

Energized

In PLCs, when the output is activated.

In electricity, when the power supply is applied to a load and there is a current flow.

Encoder

Device that changes rotation or position into a set of pulses. A value of current, voltage or a data.

Ethernet

It´s the way digital information is transmitted in Local Area Networks (LAN).

F

Ferrite Transformer

Transformer made with a core of a ferromagnetic compound.

Final Control Element

The element at the end of an action control that produces the movement, the displacement, the rotation, etc.

Forward

Rotation in one sense which is defined by you.

Function

Set of instructions that perform a complex task that can be repeated every time the function is "called" or enabled.

G

Ground

Ground is the reference point of an electrical circuit. In DC it´s the path to discharge the static and to protect the circuit. In AC it´s the reference at the transformer, where it´s connected to the neutral line.

H

Hydraulic

System that uses water or oil as fluid.

HMI

Human Machine Interface. Device between the machine and the operator to facilitate the operation and command.

I

IC

Integrated circuit. Often known as Chip.

Input

In PLCs, the connection point where the PLC receives the information from the external environment or process.

Interface

A device between two devices which acts as a communication bridge.

In PLCs it´s also applicable to operator panels.

Interlocking

The use of a contact in several parts of a electrical circuit or control program to enable or disable the operation of specific branches or rungs.

IEC 6113-3

A programming standard originated and widely used in Europe. With a growing acceptance in the rest of the world, it´s the part of the norm related to promote the standard of programming languages for PLCs.

K

Kilo

1000 units. The symbol is letter is K, when added to any unit multiply by 1000.

Example Kohm: 1000 ohms. 12KV: 12 Kilovolts= 12000 volts.

L

Ladder logic

A methodology to program PLCs. The control logic is made on rungs that resemble the steps of a ladder.

LAN

Local Area Network. A communication standard that allows multiple devices (very often computers) to share information at high speed.

LED

Light Emitting Diode. A very popular solid state device used to act as a lamp, with very low energy consumption and a long duration.

Light Beam

A very fine line of light coming from an optical Emitter and directed to an object, a mirror or to an optical Receiver. Used to protect areas.

M

Master

Applied to devices or equipment that have control on the rest.

Master Relay

A certain typical connection of a relay and relay contacts that enable or disable part of a circuit. Also used in automation for safety purposes.

Memory.

Part of the PLC where the control program or the process data is stored. Could be volatile or non- volatile according to the fact of retaining the information after power loses.

Mili

The unit divided by 1000. The symbol is the letter m. Example 20 mA : 20 Miliamperes.

Modem

Device to transmit and receive data on long distances.

Motor

Device that converts the electrical energy into mechanical energy through rotational motion.

MMI

Man Machine Interface. Same as Operator panel.

N

Noise

In electronics, unwanted signal that affects the operation of electronic equipment.

NO

Normally Open. Contact that in normal condition (without any activation) is open.

NC

Normally Closed. Contact that in normal condition (without any activation) is closed.

O

Ohm

Unit of resistance or impedance. This is

associated with loads.

Ohm´s Law

Law that express the relation among Voltage, Current and Resistance.

On

Energized. Activated. That has power applied.

Online

In PLCs , when you can have the PLC running while you are connected through a programming and simulation software.

Off

Inactive. With no power. Its power supply has been removed.

Operator

Person in charge of running a machine which is not necessarily automated.

Operator panel

Screen for the operator for him to send commands to operate the machine or visualize operation

parameters.

Optical

That uses light as the way to transport information .

Output

In PLCs, the final part of a rung. Can be external to control devices or internal to define operations (also known as marks).

P

PAC

Programmable Automation controller. New technology that combines a PLC and a PC on the same equipment.

Parallel

When each terminal of a device is connected to the same terminals on another device.

Path

Alternative route. In electrical terms, another possible way to send current to a device or output.

PC

Personal Computer.

PLC

Programmable Logic Controller.

PLC trainer

Equipment ready to be used in studying PLCs or development of automation prototypes, that includes all the needed features such as power supply,pre-wiring,push buttons, relays, fuses , etc.

Piston

The mobile part inside a pneumatic cylinder. The axis attached performs the same displacement.

Pneumatic

That uses air as the fluid to generate work.

Port.

In PLCs, special connection used for communication purposes. The port is always associated with a protocol.

Power

According to Ohm's law it´s the product of multiplying Voltage and Current.

Power Supply

Element that provides the Voltage to the circuit with the appropriate levels.

Present/Current Value

In PLCs, for timers and counter the present value is the value they have at any instant when the elements (timers and counters) are active counting.

Preset Value

In PLCs, the maximum value assigned to a counter or timer.

Protocol

Set of characters that must be sent in a certain way to be understood by another device. When the protocol is common to many devices it´s known as a "standard".

Process

A series of electrical,

chemical and mechanical operations on specific materials to produce or manufacture a product.

Pulse

Transition from Off to On and then to Off again. The duration of the ON status is very short.

Pulse Train

A continuous sequence of pulses.

Push Button

An electrical switch that is normally operated by pressing with the finger.The internal contacts it might have change its condition.

R

Rack

Metal or plastic framework that houses electronic cards, also known as modules.

RAM memory

Random Access Memory. This is a memory where the PLC temporarily stores data collected for internal operation or from the

process. Once the power is removed the data is lost.

Receiver

Device that receives the signal from an Emitter, and generates an output.

Relay

An interface device that closes or opens contacts after receiving a voltage applied to its coil.

Resistor

Electrical component that opposes to current flow.

Retentive Button

Switch that once is pressed, remains in the new state until it´s pressed again.

Reverse

Motion in opposite sense to the one selected as Forward.

S

SCADA

Supervisory Control and Data Acquisition. Software that contains all the tools to generate graphic information of industrial processes, data handling and communication protocols to collect data from different PLC manufacturers.

Sequencer

In PLCs, a sequencer is a software device that does not require ladder logic program but defining certain number of steps. It performs connection or disconnection of outputs according to the step. To jump from one step to the next, a signal from an input, a timer, a counter can be used.

Sequential Flow Chart

Program methodology used by the IEC6113 norm, that resembles the way a software programmer develops a software program, using the typical software symbols and strategy.

Series

Sort of connection where the second terminal of an element is connected to the first terminal of the next element.

Servomotor

It´s an electrical motor that uses sensing devices to feedback the position and the speed. The advantages are related to the high speed and more accurate movements that can be achieved.

Set Value

Point of reference for specific control task. Programmed value for an element that counts pulses or time.

Sensor

A device that detects or senses a signal from the external environment and converts it to a signal that can be understood by electronic equipment.

Short Circuit

A non-desirable situation that occurs when a voltage source is applied to a load with zero resistance. It´s equivalent as connecting together both terminals of the power supply.

Simulation

Imitation of a condition, when the real operation is not available or not practical at all. The perfect tools for simulations are the computers.

Solid State

Device that is based on or constructed with semiconductors. It´s the opposite to gas devices which were replaced by the inception of transistors.

Source

Element from an electrical circuit that provides the voltage or the current to the rest of the elements.

Standard signal

Range of signal values that are considered universal so, the sensor that provides standard signals can be connected to any PLC through the appropriate module.

Star

Type of connection of three loads, where the image looks like a Star of three peaks or a "Y" letter.

Step

In PLCs, one of the transition of a sequencer. During the step the programmer chose the outputs that must be active.

Stepper Motor

DC electric motor that divides the full rotation by a fixed number of steps. To operate require to be connected to a drive (Stepper Drive).

Switch

Device that breaks the continuity of a circuit, preventing the current flow.

Switching power supply

Power supply that uses a ferrite transformer which is connected and disconnected a certain given frequency. It´s more efficient that the Linear power supply.

T

Thermocouple

Sensor to measure temperature based on the junction of two different metals and the small voltage generated when the junction is submitted to a temperature.

Timer

Software element that counts time according to an internal time base.

Trainer

Person who develops training on a certain knowledge field.

Transceiver

Communication equipment that can act as a transmitter and a receiver.

Transducer

Device that changes any physical or chemical variable into an electrical signal than can be analyzed or measured.

Transmitter

Device that changes the variable from the transducer into a standard signal.

Transformer

Electromagnetic device that changes a voltage level into another one, through wire coils wrapped around a core.

Transition

Change of state. For instance from OFF to ON.

Toggle switch

Mechanical switch that manually changes from one state to another.

Touch

That is sensitive to the contact, usually you touch with a finger.

Touch Screen

Operator panel which has a screen with touch capabilities.

U

Universal Input Module

A input module that can accept any type of signal like: any type of thermocouple, any DC voltage range or any current between 0-20mA.

UPS

Uninterruptible Power Supply. Power supply that internally has a battery bank and a circuit to convert DC into AC when the main power is gone.

V

Valve

Device used to control the flow of air, liquids, oils or slurries. Can be actuated by a hand (manual valves) ,by a coil (electro-valve) , by a motor (motor valves) or by air (air actuated valves), etc.

Volts

Unit of Voltage. According to Ohm's law it´s the result of multiplying Current by Resistance.

Voltage

Amount of volts a source has. There are AC and DC Voltage sources.

W

Watt

Units of power. According to Ohm's law it's the result of multiplying Voltage by Current.

Wattage.

In electricity, the capacity of delivering or consuming power.

Winding

A coated electrical conductor forming several turns around a core that can be air, iron or any other ferromagnetic material.

Wireless

That has no physical connection with wires. The information is sent through light or radio signal.

Y

Y connection

Also known as Star connection. When connecting three loads, the connection looks like a Star of three peaks or a "Y" letter.

Z

LADDER LOGIC FROM FAMOUS PLC BRANDS

A1

Siemens®

Company Information:

Siemens is a global german company with regional offices all over the world. The Industry Automation and Drive Technologies division has three branches: Factory Automation, Process automation and Electrical equipment for buildings. The company spends more than 6,6 % of its annual profit on research and development projects.

1. Software:

Step 5, Step 7 and MicroWin.

2.PLC Families:

4. Contacts:

5. Inputs

In Siemens PLC, the inputs are named with the following format:

Ii.j

Where

    i is a number:0. 1, 2,etc.
    j is a number ranging from
    0-7,10-17 (octal numbering).

Example:

        I0.1   Input #1.
        I0.7   Input #7.
        I1.10 Input #8.
        I1.15 Input #13.

6. Coils

6.1 Internal relays

The Internal coils, marks or internal relays are denoted with the letter "C" according to the following numbering: Ci.j

6.2 Outputs

The output coils, are denoted with the letter "O" according to the following numbering: Oi.j.

Where

i is a number:0. 1, 2,etc.
j is a number ranging from 0-7,10-17 (octal numbering).

Example:

    00.1   Output #1.
    00.7   Output #7.
    01.10   Output #8.
    01.15   Output #13.

7. Timers

The PLCs from Siemens have three different types of timers: TON, TONR and TOF.

 The TON timer is a timer that starts counting time at the moment it´s enabled.

 The TOF timer is a timer that starts counting time at the moment it´s disabled.

 The TONR timer is a timer that only counts time when IN is enabled. When IN disappears the timing value remains.

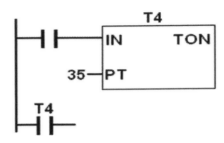

The IN input is used to enable or energize the timers. On the PT input the user can define the Preset value.

The time base depends on the timer number selected and must be consulted on the manual of the PLC used. For instance, T0 has 1ms, T1 has 10ms and T5 has 100ms of accuracy.

8. Counters

In some PLCs like the S7-200 there are 256 counters which can be named from C0 to C255. Once you have selected a counter you need to use a different number for any other type of counter you plan to use in the rest of your program.

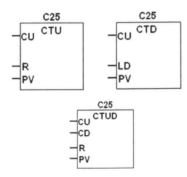

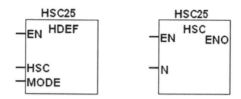

Depending on the CPU you can have up to 6 high speed counters which range from HSC0 up to HSC5.

<u>Allen Bradley</u>®

1. Company Information:

Allen Bradley is a part of the powerful conglomerate Rockwell Automation. It has a strong presence in USA and Europe. The company is one of the most important automation manufacturers in the whole world.

2. Software:

Two previous versions ,RSLogix5 and RSlogix 500, work in 16bits. The latest version is RS5000 is for controllers capable to process information in 32 bits.

3. PLC families

Micrologix, SLC.

4. Contacts:

NO   NC

5. Inputs

All the Allen Bradley's PLCs in old nomenclature name the inputs according to following format:

I1: **i / j**

Where

> **i** is a number:0. 1, 2,etc.
> **j** is a number ranging from 0-16 (Bit).
> In the new software you can use Labels so, you don't need to worry about the nomenclature.

Example:

> I1:1   Input #1.
> I1:7   Input #7.
> I1:8   Input #8.
> I1:13 Input #13.

6. Coils

6.1 Internal relays

The Internal coils or internal relays in Allen Bradley are named "Marks" They use the letters B3 according to the following:

> B3: **i**

Where

> **i** is a number:0. 1, 2,etc.

Examples:

> B3:1   Mark 1.
> B3:3   Mark 3.

7. Outputs

The Outputs always begin with the letter "O" followed by the number "0".

OO: **i**

Where :

i is a number:0. 1, 2,etc.

Some newer versions allow you to use just the letter O.

Example:

OO.1	or O:1	Output #1.
OO.7	or O:7	Output #7.
OO:10	or O:10	Output #10.
OO:15	or O:15	Output #15.

8. Timers

The PLCs from Allen Bradley have three different types of timers: ON delay (TON), OFF delay (TOF) and retentive timer (RTO).

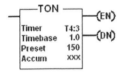

The TON timer is a timers that starts counting time at the moment it´s enabled.

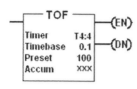

The TOF timer is a timers that starts counting time at the moment it´s disabled.

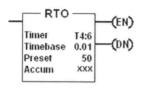

The RTO timer is a timers that counts the time only when it's enabled, without losing the time when the signal disappears.

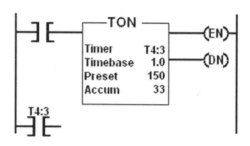

Example of connection of timer instruction.

9. Counters

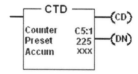

In some PLCs like the Micrologix there are 256 counters which can be named from C0 to C255. Once you have selected a counter you need to use a different number for any other type of counter you plan to use in the rest of your program.

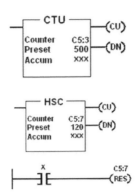

X can be any of the following:

1. A NO contact from an Input, to reset the counter manually.

2. A bit corresponding to the counter number, like the following ones:

> C5:#/13 Done. This bit´s on when the counter reaches the preset value. You can also use the following notation C5:#/DN.

Some other bits can useful when working with counters, here you will find some of them:

> C5:#/10 Update accumulator bit. This bit´s on when the accumulator of the counter has been updated. (Also C5:#/UA).
> C5:#/11 Underflow bit. This bit´s on while the counter's accumulator is below the preset value. (Also C5:#/UN).
> C5:#/12 Overflow bit. This bit´s on while the counter's accumulator is higher than the preset value. (Also C5:#/OV).

10. The Memory registers (integer).

The integer memory registers use the letters N7.

The format will be:

N7: **i**

Where **i** is a number: 0. 1, 2, etc.

General Electric GE®

1. Company Information:

General Electric (GE) and GE Fanuc are American companies working in the industrial automation field.

2. Software:

Micro.

3. PLC Families:

4.Contacts:

5. Inputs

In PLCs from GE, the inputs are named with the following format.

I j

Where:

j is a number:0000. 0001, 0002,etc.

Example:

l0001   Input #1.
l0007   Input #7.
l0008   Input #8.
l0013   Input #13.

6. Coils

6.1 Internal relays

The Internal coils or internal relays are denoted with the letter "C" according to the following numbering:

C j.

Where j is a number:0000. 0001, 0002,etc.

6.2 Outputs

The external coils or outputs are denoted with the letter "O" according to the following numbering.

O j.

Where j is a number:0000. 0001, 0002,etc.

Example:

00001   Output #1.
00007   Output #7.
00008   Output #8.
000013   Output #13.

Among the coils, there are some commands that have to be used in pairs. This means that if you use one, then the other must be used also later in the program.

> SET and RST Set and Reset. To activate an output and leave it activated.

> MCR and END. Master Control Relay. All the logic between these two instructions is only considered if the MCR command is active.

> SKIP and END. Skips the rungs until the program find END.

7. Timers

The PLCs from GE have two types of timers: ON TIMER and OFF TIMER.

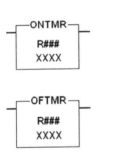

The ONTMR timer is a timer that starts counting time at the moment it's enabled.

The OFTMR timer is a timer that starts counting time at the moment it's disabled.

The time base is of 1ms of accuracy.

8. Counters

Counters which can be named from C0 to C255. Once you have selected a counter you need to use a different number for any other type of counter you plan to use in the rest of your program.

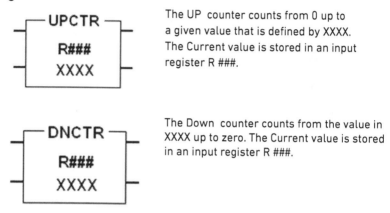

The UP counter counts from 0 up to a given value that is defined by XXXX. The Current value is stored in an input register R ###.

The Down counter counts from the value in XXXX up to zero. The Current value is stored in an input register R ###.

XXX can be a register as well. The Registers can be selected from R1 up to R500.

Automation Direct®

1. Company Information

It started as Controles Direct. A company that could sell PLCs without the overheads costs of big companies. All those savings were translated to the products allowing to sell very low cost, yet powerful PLCs. Currently, the company sells all types of automation products through the online company and the Valuable Added Resellers (VARs).

2. Software:

Click, Directsoft5.3.

3. PLC Families:

Click, DL05/06, DL130, DL205, DL305, DL405.

4. Contacts:

5. Inputs

In Automation direct PLCs, the inputs are named with the following format:

Xj

Where:

j is a number ranging from.

0-7,10-17 (octal numbering).

Example:

X0   Input #1.
X7   Input #8.
X10   Input #9.
X14   Input #13.

6. Coils

6.1 Internal relays

The Internal coils or internal relays are denoted with the letter "C" according to the following numbering: Ci

Where:

i is a number ranging from

0-7,10-17 (octal numbering).

6.2 Outputs

The outputs are identified by a "Y" and use the following format:

Yi .

Where

i is a number ranging from
0-7,10-17 (octal numbering).

Example:

YO   Output #1.
Y6   Output #7.
Y7   Output #8.
Y14   Output#13.

7. Timers

The PLCs from Automation Direct have four different types of timers: TMR, TMRF, TMRA, TMRAF.

The TMR timer is a timer that starts counting time at the moment it´s enabled. The Time base is 0.1sec. The preset value is 30 so the timing is for 3 seconds.

The TMRA timer is a timer that counts the time only when it's enabled, without losing the time when is this signal disappears. When enabled again it continues from the last value. The Time base is 0.1sec. The preset value is 30 so the timing is for 3 seconds.

The TMRF timer is a timer that counts the time only when it's enabled. The Time base is 0.01sec. The preset value is 30 so the timing is for 0,3 seconds,

The TMRAF timer is a timer that counts the time only when it's enabled, without losing the time when is this signal disappears. When enabled again it continues from the last value. The Time base is 0.01sec. The Preset value is 30 so the timing is for 0.3 seconds.

The Preset value in the examples on the left is a constant but can be replaced by a memory register that allows to program this value from another device like tan operator panel.

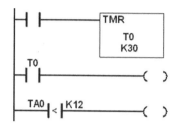

Example of connection of timer instruction.

The current value for a given timer can be accessed asking for the TAX, where X is the timer number. For example a 10 seconds timer named T3 , will have the timing in real time stored in register TA3.

8. Counters

Counters in Automation Direct PLCs are define as CTX. Where X is a number with octal numbering (0-7, 10-17, 20-27, etc). Once you have selected a counter you need to use a different number for any other type of counter you plan to use in the rest of your program. There are two types of counters. The normal ones and the up/down counter.

```
 H |----------------------[CNT]
  CT4                      CT4
 H |                       K345

  CT4
 H |------------------------( )

  CTA4 , |K120
 |----| > |------------------( )
```

Example of how to use a counter.

In this case a contact from the same counter is used to Reset the counter and then to activate an output in the next rung. You can ask for the counter value at any time and use this information to make decisions as in the last rung.

Memory registers:

The memory registers are denoted with the letter V and , depending from the PLC range from V1400 to V1777 in octal. If you use a memory position for any of the elements of the PLC, somewhere in the program you need to use some instruction to store a value on that specific memory register.

The counter CT4 counts from 0 up to the Preset value, which in this case is 345. The counter has two inputs, the upper one is the pulse input, the lower one is to Reset the counter.

The UPDOWN counter CT21 is a counter that starts counting Up if it receives pulses in the upper input. It counts down if receives pulses in the middle input. The lower input is the reset input.

Triangle Research Lmtd®

1. Company Information:

This company started in 1993 in Singapore. They have designed very powerful, yet low cost, PLC equipment. Two major advantages, their PLCs are OEM oriented and loaded with an arsenal of features that are very difficult and expensive to find in the market of PLC manufacturers.

2. Software:

Wintrilogi®, Trilogi®.

3. PLC Families:

F-series, M-series can be programmed in Ladder + Basic.
H-series and E10 series can be programmed in pure ladder.

4. Contacts:

$$\dashv\vdash \quad \dashv\!/\!\vdash$$

NO   NC

5.Inputs

In these PLCs, the inputs are named with no particular format, they can use the label or name directly, prior definition in the I/O table in the Inputs table.

The table has numbering so, the Inputs have to be assigned.

first before using them.

The numbering is consecutive ranging from 0-255.

Example:

>Start   in Input #1.
>Stop   in Input #7.
>Max_level in   Input #8.
>Reset in   Input #13.

6. Coils

6.1 Internal relays

The Internal coils or internal relays have their own instruction. They vary in number , according to the PLC series. There is a table for Internal Relays in the I/O table. They can have meaningful names.

6.2 Outputs

The outputs must be created first in the Output table of the I/O table. They can have a name of 8 characters long to make it more understandable. The number of outputs depends on the PLC model.

Examples:

>Motor   in Output #1.
>Lamp   in Output #7.

Alarm   in Output #8.
Ejector in Output #13.

7. Timers

Since Triangle Research has different PLC series, the number of timers depends on the family and can be defined in the Timers table in IO table.

The time base is 0.1 seconds for all the timers. Since the M and F series can be programmed in BASIC language, there is an easy way to access the Set value and the Present Value of any timer used in the ladder logic. It has some instructions that allow to perform data manipulation with the timing parameters.

The timer can be used as a normal ladder instruction (E10 and H series) or used inside a Customer function which is written in Basic language (M and F series).

SetTimerSV: To program the SET VALUE of a given timer at any time.

GetTimerSV: To ask for the SET VALUE of a given timer at any time of the program execution.

TimerPV[n]: To ask for the PRESENT VALUE of a given timer at any time.

DELAY: To generate a small delay when running a small program in Basic in a Customer function.

Examples:

>   SetTimerSV 1,2000.
>   Sets the timer 1 for 200 sec.
>   SetTimerSV 2, GetTimerSV(1)+10.

Sets the timer 2 with the value of another timer (the number 1) and add the value of 10 (1sec).

8. Counters

The counters are created in the Counters table in the IO tables. E10 and H series have less counters than more sophisticated series like M and F.

The Counter can be used as a normal ladder instruction (E10 and H series) or used inside a Customer function which is written in Basic language (M and F series).

>   SetCtrSV: To program the SET VALUE of a given counter at any time.

>   GetCtrSV( ) To ask for the SET VALUE of a given counter at any time of the program execution.

>   CtrPV[n] To ask for the PRESENT VALUE of a given counter at any time.

Examples:

>   SetCtrSV 3,2489.
>   Sets the Counter 3 to count up to 2489.

>   SetCtrSV 4, GetCtrSV(1)+15.
>   Sets the Counter 4 with the value of another counter (the number 1) and add the value of 15.

We do appreciate your help!

PLC Brand_____

Company Information:

Software:

PLC Families:

Contacts:

Inputs

In these PLCs, the inputs are named

Example:

Coils

6.1 Internal relays

6.2 Outputs

Timers

Counters

The counters are

Examples:

FUNDAMENTALS ON ELECTRICITY

A2

If you don´t know much about electricity, this is a good start. You will learn about the basic jargon in the electrical world.

Appendix

Fundamentals on Electricity

In the World of electric signals two main systems can be considered: Direct Current (DC) and Alternating Current (AC).

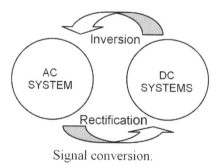

Signal conversion.

The process on generating AC from DC is called Inversion and the devices that do this job are called Inverters.

On the contrary, to generate DC from AC you use a device called Rectifier and the process is called of Rectifying.

To have a better understanding of the AC and DC systems, let's assume that we can make a graphic of any electrical signal through time.

The DC signal has the same polarity all the time. It can be either negative or positive all the time. When it's changing polarity at any given time it becomes an AC signal. So, for definition: An AC signal is a signal that changes polarity periodically. The speed of these polarity changes is called frequency.

You can draw two axis represented by arrows (positive and negative) to show a signal thru time.

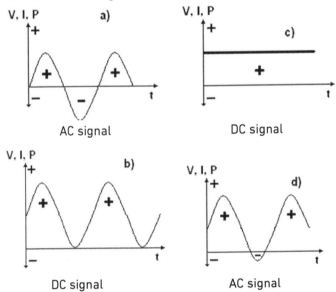

AC signal

DC signal

DC signal

AC signal

From the diagrams shown on the previous page you can conclude that a) and d) are AC signals and that b) and c) are DC signals. A Negative signal is under the line of time. A Positive signal is above the line of time.

Equipment with capabilities of providing energy through AC or DC signals is called Source. Devices that receive the energy from sources are called Loads.

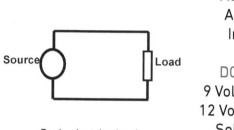

Basic electric circuit.

AC Sources:
AC outlets.
Inverters.
UPSs.
DC Sources:
9 Volts Batteries.
12 Volts Batteries.
Solar panels.

Power supply Symbols:

DC Sources.

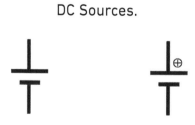

AC sources.

The electrical circuit's a system formed by connecting a Source to a Load, for the energy interchange.

The Source provides the energy and the Load converts it to work, movement, heat, etc.

DC Source with load.

AC Source with load.

All the electric circuits are governed by the Ohm's Law.

"If a Source is connected to a Load, the result is an electron flow called current flowing through the circuit".

This current can be calculated by the following formula:

I=V/R

I= Current of the circuit in Amps.
V= Voltage source in Volts.
R= Resistance of load in Ohms.

$$I = \frac{V}{R} \qquad \text{Ohm's law}$$

A very good way to learn and remember the Ohm's law is to use the following triangle:

$$I = \frac{V}{R}$$

The idea is that you can calculate any given parameter based on the remaining two.

$$V = I \times R \qquad R = \frac{V}{I}$$

If you need to know the value of any specific parameter, just use your finger to cover it on the triangle. The other uncovered two parameters are a formula that uses the parameters in the exact way they are located: One parameter divided by the other or Current times Resistance if you look specifically for Voltage.

You can see that there is a current only if both terminals of the Source are connected to the corresponding terminals of the Load. In order to control the connection we will introduce a new element named the Switch.

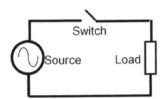

Three elements circuit.

We have current flowing through the circuit only when the switch is closed. When this situation occurs you can think that the switch is not present at all.

The three elements (Sources, Loads and Switches) can be connected in the following different ways:

Series connection.

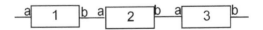

Connecting three elements in series.

One end of the element 1 is connected to one end of the element 2. The other end of element 2 is connected to one end of element 3 and so on.

Parallel connection.

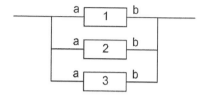

Connecting three elements in parallel.

All the "a" terminals are connected together. The same applies to the "b" terminals.

The series and parallel connection can have as many elements as you want. You can make combinations as well.

In some cases you can make connections without paying attention to terminals, but in some others, based on the nature of the element, the resultant electrical circuit's affected by the way you make this connections. In this case these elements are considered as devices with bias or polarity.

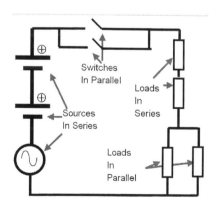

A more complex circuit with elements connected in different ways.

An electrical circuit can be simplified.

In automation, some standards are widely used:

DC Sources:

12, 24 DC V(DCV stands for DC Volts)with + and − terminals.

AC Sources: 120, 240, 440ACV (ACV stands for AC Volts).

The AC sources use the following names for their lines or terminals.

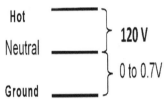

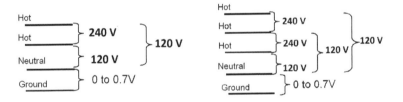

The Hot wires are also called Phases.

When the load is connected between the Hot and Neutral or between the Hot and Hot you can say that the electric circuit´s Monophasic or Single phase.

When a device or load is simultaneously connected to three different Hot wires this is called a Tri-phase circuit.

The Neutral and Ground cables are independent cables in normal installations, but they are connected together before leaving the equipment that acts as the source: The Transformer.

To recognize the terminals on an AC outlet:

The standard voltage of any AC outlet is 120ACV in America and 240ACV in Europe.

For DC sources, it's assumed that the Red color is + and Black color is -, but always check this since this convention is different in China.

The voltage sources in the ideal condition can supply as much current as required when connected to a load. The truth is that in real conditions the maximum current is limited by several parameters: construction, wire gauges, fuse or breaker protection, etc.

There is a way to relate the Voltage of the source and the maximum current it can provide. This is what we call Power.

The ohm's law for power can defined as follows:

Power in Watts is the result of multiplying the Voltage times the Current. For a source it's the voltage of the source times the current drained by the Load.

For practical purposes you can use the Triangle to get any parameter based on the other two or simply use the formulas below, where:

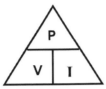

I = Current of the circuit in Amps.
V = Voltage source in Volts.
P = Power of the source in Watts.

$$P = V \times I \qquad\qquad V = \frac{P}{I}$$

$$I = \frac{P}{V}$$

PLC TRAINING
STATIONS

Affordable and powerful equipment for hands-on learning and training on Ladder Logic and PLCs.

If you look for affordable and powerful trainers, that allow you to learn and practice, this is for sure your best choice. If you plan to develop automation projects, these trainers will help you in having your prototypes up and running in no time.

More New Technologies:
You can learn the classical concepts or you can innovate with new approaches in the changing world of industrial automation.

Extremely Low Cost:
For the cost of a PLC you can buy a complete trainer with pushbuttons, switches, relays, LEDs and power supply...

Not only Educational But Real:
These PLCs are used f or m any multinationals. Anyone or any educational center can take his application to the real world.

More Control Techniques:
You can learn more technologies besides the classical ladder programming such as BASIC.

More New Topics:
Could you imagine monitoring and controlling through EXCEL®? You can also operate our PLCs over the Internet. Try robotics and telemetry.

More Learning Levels:
Learners can begin to work with smaller, less complex and lower cost PLCs before using higher or more expensive PLC technologies.

If you are an EDUCATOR looking for the best way to provide a course on Programmable Logic Controllers PLCs this methodology is quite recommended.

The world of automation is in constant change. Most of the industries have or use more than one automation brand. This is the good part of technology: some manufacturers can have really good products for one automation branch but can be overpassed by some others in other fields and applications.

Our low cost PLC training equipment is designed to cover most of the PLC brands so you learn the way education should be:

With no particular PLC preferences at all.

•Affordable costs, allowing you to acquire more training units with limited budgets. Our suggestion is that you have one or two students per PLC trainer.

•We try to cover more topics. Your students can do better and more sophisticated projects.

•You can allow the students to install the programming software in their PCs or Laptops and, since it has a complete simulation. They can work at home in practicing and studying by themselves or working on the homework you have previously assigned.

•You can impart more hands-on experience. Every class can be a different project that can be downloaded to the PLC trainer.

Used by:
Australian RMIT, TAFE y Old Dominion University, Purdue University, State University of New York, Sait Polytechnic (Canada), Torreon University (Mexico), Kent State University, Florida International University, Miami Dade College and many more all over the world.

Latin-Tech

All the models include a Software Simulator

PTS E10

A good way to start learning about PLCs. Six inputs/ four outputs with capabilities to connect and control real equipment.

PTS NANO

For second steps and knowledge, this is the lowest cost trainer has it all: analog, digital, Ethernet, Ladder+ Basic.

PTS F888

A basic super PLC trainer with more features: analog and digital inputs and outputs, Ethernet, Serial ports, display. Ladder + Basic.

PTS F1616

A super PLC trainer with many more features: more analog and digital inputs and outputs, Ethernet, three serial ports, Infrared port, display. Ladder + Basic.

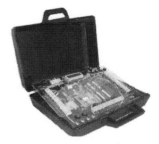

Visit **www.lt-automation.com** for more information.

Coming soon....Trainers for other brands

Allen Bradley *Allen-Bradley*

PTS 10BXB
A brand which is recognized as one of the world´s leader. This trainer helps you to understand the basics about programming an Allen Bradley´s PLC.

Automation Direct

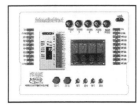

PTS CLICK
A fast growing brand because of the innovative features in programming and the excellent cost. A mini PLC trainer with only digital inputs and outputs.

PTS CLICK-A
A new version of the brand with digital and analog inputs and outputs. A perfect balance between cost and features.

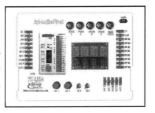

Siemens SIEMENS

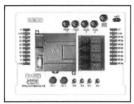

PTS S7-1200S
This famous German brand is a proven technology which is used on several industrial processes all over the world. This trainer is based on its newest PLC.

Attention!!! Educational institutions.

Most of the educational institutions are interested in providing the best educational tools to guarantee the right background that new professionals deserve. But, with the current budget restrictions the task of spending the money wisely becomes really difficult.

Comments from Associated Professor Ambrose G Barry, Eng. Tech Dept, UNCC

In reference to the proposal from Latin-tech, Inc, it should be noted that they are the only company providing ALL of the components required, and ALL of the capabilities required for the PLC trainers.

They are the only company providing PLCs with Ethernet & Internet connectivity at zero cost. Additionally the system, AS PROVIDED BY Latin Tech, are programmable in normal ladder logic, in BASIC, in C, and their unique software allows control & monitoring thru Microsoft EXCEL spreadsheets.

Their units are very popular both in industry (for use as PLCs and as trainers) and at many universities, including Old Dominion, Purdue, Kent State, State Univ of NY, and others.

A significant amount of time was spent researching this item, and it's felt that the systems as provided by Latin Tech, will not only completely satisfy our educational needs, but that they are by far the most economical systems providing all of the features available, including the items mentioned above."

Comments from Professor John Hackword Old Dominion University (www.eng.odu.edu)

"I use TRiLOGI in the college classroom. I have each student download and install TRiLOGI on their home computer, and I then give them programming assignments to do at home. They individually write, run and debug their PLC programs, and then email their program files to me for testing and grading. Since TRiLOGI will simulate PLC operation off-line, each student is able to obtain hands-on programming and debugging experience without having to wait for their turn on an expensive laboratory PLC. The students enjoy using TRiLOGI because they can see their programs run at home without loading them into an actual PLC, it's easy to learn, and the cost for the off-line version is zero."

Why Purdue University is one of the most important
engineering schools of the world?

This is IE575 course, conducted at Purdue University. Our
equipments were selected to teach students about industrial
automation. Some project pages of the link contain photos of
projects constructed using our Super PLC. You can check this
link:http://cobweb.ecn.purdue.edu/~ie575/

If you want to learn more....buy our SECOND BOOK

Fundamentals on Programmable Logic Controllers
Intermediate Level

Preliminary Contents

From the author
About the author
Introduction
1. Analog signals.
2. Codes and conversion codes.
3. Analog variables.
4. Analog inputs.
5. Analog outputs.
6. Ladder logic and Tbasic.
7. Data manipulation.
8. Communication and protocols.
9. Programming and simulation
   software trilogi.
10. Writing a program.
11. Tbasic.
12. Displaying information.
13. 40 Lessons.
14. 10 Projects.

Includes
SOFTWARE
SIMULATOR

Theory
40 Step by Step Lessons
8 Real and Practical Projects
Software Simulator

Basic Instructions for several PLCs:
Allen Bradley®, Siemens®, Automation Direct®, General Electric®
and Triangle Research®

NEW!!! Real Pilot Plants!

A new revolution for education on PLCs, control, instrumentation and automation courses for every engineering and technical institution.

Soda/Malt/Beer pilot plant

For the very first time you can have SEVERAL groups of students simultaneously attending a lab practice for level, flow, pressure or temperature.

They learn and practice about the concepts and at the end of the course, they prepare a batch of the product which later they can try and drink. The kind of lab and education nobody forgets.

Biofuel pilot plant

Learn about the technology that intends to replace fossil fuels. Produce Ethanol from biomass using local vegetable resources. Explore all the alternatives while learning about all the automation processes.

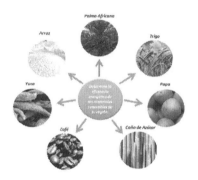

Algae´s biofuel pilot plant

This technology is one of the best alternatives to replace fossil fuels since it doesn´t affect food sustainability. Produce Ethanol from Algae and lean about all the involved automation processes.

Visit **www.lt-automation.com** for more information.

INDEX